T0227768

RSS-AoA-based Target Localization and Tracking in Wireless Sensor Networks

RIVER PUBLISHERS SERIES IN COMMUNICATIONS

Series Editors

ABBAS JAMALIPOUR
The University of Sydney
Australia

MARINA RUGGIERI
University of Rome Tor Vergata
Italy

JUNSHAN ZHANG
Arizona State University
USA

Indexing: All books published in this series are submitted to Thomson Reuters Book Citation Index (BkCI), CrossRef and to Google Scholar.

The "River Publishers Series in Communications" is a series of comprehensive academic and professional books which focus on communication and network systems. The series focuses on topics ranging from the theory and use of systems involving all terminals, computers, and information processors; wired and wireless networks; and network layouts, protocols, architectures, and implementations. Furthermore, developments toward new market demands in systems, products, and technologies such as personal communications services, multimedia systems, enterprise networks, and optical communications systems are also covered.

Books published in the series include research monographs, edited volumes, handbooks and textbooks. The books provide professionals, researchers, educators, and advanced students in the field with an invaluable insight into the latest research and developments.

Topics covered in the series include, but are by no means restricted to the following:

- Wireless Communications
- Networks
- Security
- Antennas & Propagation
- Microwaves
- Software Defined Radio

For a list of other books in this series, visit www.riverpublishers.com

RSS-AoA-based Target Localization and Tracking in Wireless Sensor Networks

Slavisa Tomic

Universidade de Lisboa
Portugal

Marko Beko

Universidade Lusófona, and CTS/UNINOVA
Portugal

Rui Dinis

Instituto de Telecomunicações, and FCT/UNL
Portugal

Milan Tuba

John Naisbitt University
Serbia

Nebojsa Bacanin

John Naisbitt University
Serbia

River Publishers

Routledge
Taylor & Francis Group
LONDON AND NEW YORK

Published 2017 by River Publishers
River Publishers
Alsbjergvej 10, 9260 Gistrup, Denmark
www.riverpublishers.com

Distributed exclusively by Routledge
4 Park Square, Milton Park, Abingdon, Oxon OX14 4RN
605 Third Avenue, New York, NY 10017, USA

First issued in paperback 2023

RSS-AoA-based Target Localization and Tracking in Wireless Sensor Networks / by Slavisa Tomic, Marko Beko, Rui Dinis, Milan Tuba, Nebojsa Bacanin.

Publisher's Note
The publisher has gone to great lengths to ensure the quality of this reprint but points out that some imperfections in the original copies may be apparent.

ISBN 13: 978-87-7022-980-7 (pbk)
ISBN 13: 978-87-93519-88-6 (hbk)
ISBN 13: 978-1-003-33933-5 (ebk)

Contents

Preface

The desire for precise knowledge about the location of a moving object at any time instant has motivated a great deal of scientific research recently. This is owing to a steady expansion of the range of enabling devices and technologies, as well as the need for seamless solutions for location-based services. Besides localization accuracy, a common requirement for emerging solutions is that they are cost-abstemious, both in terms of the financial and computational cost. Hence, development of localization strategies from already deployed technologies, e.g., from different terrestrial radio frequency sources is of great practical interest. Amongst other, these include localization strategies based on received signal strength (RSS), time of arrival, angle of arrival (AoA) or a combination of them.

RSS-AoA-based Target Localization and Tracking in Wireless Sensor Networks presents recent advances in developing algorithms for target localization and tracking, reflecting the state-of-the-art algorithms and research achievements in target localization and tracking based on hybrid (RSS-AoA) measurements.

Technical topics discussed in the book include:

Centralized RSS-AoA-based Target Localization
Distributed RSS-AoA-based Target Localization
RSS-AoA-based Target Tracking via Maximum A Posteriori Estimator
RSS-AoA-based Target Tracking via Kalman Filter
RSS-AoA-based via Sensor Navigation

This book is of interest for personnel in telecommunications and surveillance industries, military, smart systems, as well as academic staff and postgraduate/research students in telecommunications, signal processing, and non-smooth and convex optimization.

Acknowledgments

This work was partially supported by Fundação para a Ciência e a Tecnologia under Project PEst-OE/EEI/UI0066/2014 (UNINOVA) and Project UID/EEA/50008/2013 (Instituto de Telecomunicações), Program Investigador FCT under Grant IF/00325/2015 and Grant SFRH/BD/91126/2012. M. Tuba is supported by the Ministry of Education, Science and Technological Development of Republic of Serbia, Grant No. III-44006.

List of Figures

List of Tables

List of Abbreviations

AoA	Angle of arrival
BS	Base station
CRB	Cramer-Rao lower bound
FIM	Fisher information matrix
GPS	Global positioning system
GTRS	Generalized trust region sub-problem
KF	Kalman filter
LAN	Local area network
LE	Localization error
LoS	Line-of-sight
LS	Least squares
MAC	Medium access control
MAP	Maximum *a posteriori*
MDS	Multidimensional scaling
MEMS	Micro-electro-mechanical systems
ML	Maximum likelihood
NLoS	Non-line-of-sight
NRMSE	Normalized root mean square error
PDF	Probability density function
PF	Particle filter
PLE	Path loss exponent
RMSE	Root mean square error
RF	Radio frequency
RSS	Received signal strength
RSSD	Received signal strength difference
RTT	Round-trip time
SDP	Semidefinite programming
SOCP	Second-order cone programming
SR	Squared range
TDoA	Time-difference of arrival
SoA	State of the art

STD	Standard deviation
ToA	Time of arrival
ToF	Time of flight
UT	Unscented transformation
WLAN	Wireless local area network
WLS	Weighted least squares
WSN	Wireless sensor network

Nomenclature

For reference purposes, some of the most common symbols used throughout the work are listed below. Throughout the work, upper-case bold type, lower-case bold type and regular type are used for matrices, vectors and scalars, respectively.

$\mathbb{R}$	the set of real numbers
$\mathbb{R}^n$	n-dimensional real vectors
$\mathbb{R}^{m\times n}$	$m \times n$ real matrices
$[\boldsymbol{A}]_{ij}$	the ij-th element of $\boldsymbol{A}$
$\boldsymbol{A}^T$	the transpose of $\boldsymbol{A}$
$\boldsymbol{A}^{-1}$	the inverse of $\boldsymbol{A}$
$\text{tr}(\boldsymbol{A})$	the trace of $\boldsymbol{A}$
$\boldsymbol{A} \succeq \boldsymbol{B}$	the matrix $\boldsymbol{A} - \boldsymbol{B}$ is positive semidefinite
$\boldsymbol{A} \otimes \boldsymbol{B}$	the Kronecker product of $\boldsymbol{A}$ and $\boldsymbol{A}$
$\boldsymbol{I}_n$	the $n \times n$ identity matrix
$\mathbf{0}_{m\times n}$	the $m \times n$ matrix of all zero entries
$\mathbf{1}_n$	the n-dimensional column vector with all entries equal to one
$\boldsymbol{e}_i$	the i-column of an identity matrix
$\text{diag}(\boldsymbol{x})$	the square diagonal matrix with the elements of vector $\boldsymbol{x}$ as its main diagonal, and zero elements outside the main diagonal
$\lVert\boldsymbol{x}\rVert$	the Euclidean norm of vector $\boldsymbol{x}$; $\lVert\boldsymbol{x}\rVert = \sqrt{\boldsymbol{x}^T\boldsymbol{x}}$, where $\boldsymbol{x} \in \mathbb{R}^n$ is a column vector
$p(\cdot)$	probability density function
$\mathcal{N}(\boldsymbol{\mu}, \boldsymbol{\Sigma})$	real-valued Gaussian distribution with mean vector $\boldsymbol{\mu}$ and covariance matrix $\boldsymbol{\Sigma}$
$\sim$	distributed according to
$\approx$	approximately equal to
$\log_a(x)$	the base-a logarithm of x; when a is omitted it denotes the natural algorithm

Abstract

This work addresses target localization problem in wireless sensor networks (WSNs) by taking advantage of combined, received signal strength (RSS)-angle of arrival (AoA), measurements. Both non-cooperative and cooperative WSNs are investigated for different settings of the localization problem. We start by formulating maximum likelihood (ML) estimation problem. The general idea is to tightly approximate this estimator by another one whose global solution is a *close* representation of the ML solution, but is easily obtained due to greater smoothness of the derived objective function. More specifically, in the case of centralized localization, a weighted least squares (WLS) estimator is developed by converting the non-cooperative localization problem into a generalized trust region sub-problem (GTRS) framework, and a semidefinite programming (SDP) estimator is derived by applying SDP relaxations to the cooperative localization problem. Furthermore, a distributed second-order cone programming (SOCP) estimator is developed, and an extension of the centralized WLS estimator for non-cooperative WSNs to distributed conduction in cooperative WSNs is also presented. The second part of this work is committed to RSS-AoA target tracking problem. Both settings with fixed/static anchors and mobile sensors are investigated. First, the non-linear measurement model is *linearized* by applying Cartesian to polar coordinates conversion. Prior information extracted from target state model is integrated with the derived model, and a maximum *a posteriori* (MAP) algorithm is developed. Moreover, by taking advantage of the derived model and the prior knowledge, Kalman filter (KF) algorithm is designed. Finally, we allow sensors mobility, and describe a simple navigation routine to manage their movement in order to further reduce the estimation error.

The described algorithms are assessed and validated through simulation results and real indoor measurements.

1

Introduction

1.1 Motivation

Wireless sensor network (WSN) generally refers to a wireless communication network which is composed of a number of devices, called sensors, allocated over a monitored region in order to measure some local quantity of interest [1]. Due to their autonomy in terms of human interaction and low device costs, WSNs find application in various areas, like event detection (fires, floods, hailstorms) [2], monitoring (industrial, agricultural, health care, environmental) [3, 4], energy-efficient routing [5], exploration (deep water, underground, outer space) [6], and surveillance [7] to name a few. Recent advances in radio frequency (RF) and micro-electro-mechanical systems (MEMS) permit the use of large-scale networks with hundreds or thousands of nodes [1].

In many practical applications (such as search and rescue, target tracking and detection, cooperative sensing and many more), data acquired inside a WSN are only relevant if the referred location is known. Moreover, accurate localization of people and objects in both indoor and outdoor environments enables new applications in emergency and commercial services (*e.g.* location-aware vehicles [8], asset management in warehouses [9], navigation [10–13], *etc.*) that can improve safety and efficiency in everyday life, since each individual device in the network can respond faster and *better* to the changes in the environment [14]. Therefore, accurate information about sensors' locations is a valuable resource, which offers additional knowledge to the user.

However, sensors are small, low cost and low power nodes commonly deployed in a large number over a region of interest with limited to non-existing control of their location in space, *e.g.* thrown out of an aeroplane for sensing in hostile environments [15]. Besides sensing, sensors have a limited (due to their battery life) capability of communicating and processing the acquired data. Installing a global positioning system (GPS) receiver in each sensor is a possible solution, but it would severely augment the network costs and restrict its applicability [16]. Besides, GPS is ineffective indoor, dense urban and forest environments or canyons [17]. In order to maintain low implementation costs, only a small fraction of sensors are equipped with GPS receivers (called anchors), while the remaining ones (called targets) determine their locations by using a kind of localization scheme that takes advantage of the known anchor locations [18]. Since the sensors have minimal processing capabilities, the key requirement is to develop localization algorithms that are fast, scalable and abstemious in their computational and communication requirements. Also, making use of existing technologies (such as terrestrial RF sources) when providing a solution to the object localization problem is strongly encouraged. Nevertheless, WSNs are subject to changes in topology (*e.g.* node mobility, adding nodes, node and/or link failures), which aggravates the development of even the simplest algorithms.

The idea of wireless positioning was initially conceived for cellular networks, since it invokes many innovative applications and services for its users. Nowadays, rapid increase of heterogeneous smart-devices (mobile phones, tablets) which offer self-sustained applications and seamless interfaces to various wireless networks is pushing the role of the location information to become a crucial component for mobile context-aware applications [18]. Even though we limit our discussion to sensor localization in WSN here, it is worth noting that, in practice, a base station (BS) or an access point in local area network (LAN) can be considered as an anchor, while other devices such as cell phones, laptops, tags, *etc.*, can be considered as targets.

1.2 Localization Schemes

Nowadays, RF signals come from a wide variety of sources and technologies, and they can be used for localization purpose. Location information can be obtained by range-based or range-free measurements. Here, the focus is on the former ones exclusively, since they provide higher estimation accuracy in general. Hence, the locations of the targets in a WSN are determined by using a kind of localization scheme that relies on the known locations of the anchors and range measurements between targets and anchors. Range measurements can be extracted from different characteristics of the radio signal, such as time of arrival (ToA) [19], time-difference of arrival (TDoA) [20], round-trip time (RTT), time of flight (ToF) [21], angle of arrival (AoA) [22] or received signal strength (RSS) [23, 24], depending on the available hardware. The trade-off between the localization accuracy and the implementation complexity of each technique is a very important factor when deciding which method to employ. For example, localization based on ToA or TDoA (including the GPS) gives high estimation accuracy, but it requires a very complex process of timing and synchronization, thus making the localization cost-expensive [25]. Although less accurate than the localization using ToA, TDoA or AoA information, localization based on the RSS measurements requires no specialized hardware, less processing and communication (and consequently, lower energy), thus making it an attractive low-cost solution for the localization problem [1, 16]. Another attractive low-cost approach might be exploiting RTT measurements, which are easily obtained in wireless local area network (WLAN) systems by using a simple device such as a printed circuit board [26]. Even though RTT systems circumvent the problem of clock synchronization between nodes, the major drawback of this approach is the need for double signal transmission in order to perform a single measurement [27].

Recently, hybrid systems that fuse two measurements of the radio signal have been investigated [26, 28–38]. Hybrid systems profit by exploiting the benefits of combined measurements (more available information), taking advantage of the strongest points of each technique and minimizing their drawbacks. On the other hand, the price to pay for using such systems is the increased complexity of network devices, which increases the network implementation costs [1, 16].

In order to acquire the necessary measurements, node communication is required, which can be non-cooperative or cooperative [18]. The former one allows targets to communicate with the anchors exclusively, Figure 1.1(a), while the latter one allows targets to communicate with all sensors inside their communication range, whether they are anchors or targets, Figure 1.1(b).

Typically, data processing inherent to localization schemes can be performed in a centralized (network-centric positioning) or a distributed (self-positioning) fashion [18]. On the one hand, existence of a central processor (sensor or a BS) is required for the former approach. Central processor gathers all measurements through wireless transmissions and produces a map of the entire network [30–33]. This approach

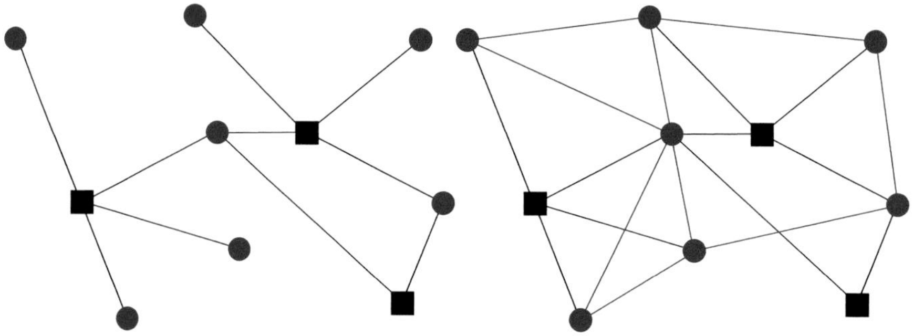

(a) Non-cooperative WSN (b) Cooperative WSN

Figure 1.1 Example of a WSN with three anchors (black squares) and seven targets (blue circles).

is characterized by fundamental optimality and stability [18]. However, in large-scale networks, a high energy drain is likely to occur at and near the central processor, caused by bottlenecks [16]. Likewise, computational complexity of a centralized approach depends highly on the network size. In many applications a central processor (or one with enough computational capacity) is not available. Furthermore, confidentiality may prevent sharing objective functions between sensors in some practical applications [39]. On the other hand, the latter approach is distinguished by low computational complexity and high-scalability, which makes it a preferable solution for large-scale and highly-dense networks [18]. However, distributed algorithms are executed iteratively, which makes them sensitive to error propagation and raises energy consumption. When determining which approach to use for a given application, one has to take into consideration all of the above properties, but if often comes down to efficiency comparison in terms of energy consumtpion. In general, when the average number of hops to the central processor is higher than the necessary number of iterations required for convergence, the distributed approach is likely to be more energy-efficient and vice versa [1].

1.2.1 Overview of Localization Techniques

A detailed survey on localization algorithms can be found in [40], and a brief overview of the state of the art (SoA) related with each chapter's discussion will be provided at the beginning of each chapter. Here, a general overview of the most commonly used localization techniques is presented.

1.2.1.1 Range-free localization

The most commonly used range-free localization technique is fingerprinting. Generally, it can be described as a multiple hypothesis testing decision problem, where the objective is to deduce the best hypothesis (location of the target) based on previously acquired observations, *i.e.*, fingerprints. In practice, a fingerprinting localization method requires two phases: the training and the localization phase. During the training phase, fingerprints are collected at all sample locations [41]. During the localization phase, an obtained radio measurement is compared with all observations collected at sample locations, and the best fit sample location is taken as the estimated target location.

On the one hand, the main advantage of this technique is the flexibility to any radio interface. On the other hand, the localization accuracy depends on the reliability (quantity and up-to-date) of the training data, the error in the synthesis of the fingerprint parameters, and the sensitivity of the algorithm to changes of the environment.

To improve the robustness of the location estimation with respect to the inaccuracy of training data, several techniques are proposed in the literature. For instance, in [42] statistical learning is used to design an algorithm based on support vector machine.

1.2.1.2 Range-based localization

Range-based localization technique is widely used nowadays owing to its potentially high accuracy, applicability to different radio technologies and ease of implementation. Within this approach, one can distinguish between range and range-difference based methods. Some of the most popular range-based localization techniques are briefly described in the following text.

Geometric-based Techniques. In the case where the noise is absent and the number of anchors is low, geometric-based techniques are appealing owing to their simplicity. Some basic and intuitive geometric methods are trilateration, triangulation and multilateration. Trilateration technique makes use of distance measurement and known location of anchor to describe a circle around the anchor with the radius equal to distance measurement [43]. Then, by using at least three anchors in

2-dimensional space, it locates the target by calculating the intersection of the circles based on simultaneous range measurements from anchors, Figure 1.2(a). Triangulation is used when the direction of target instead of the distance is estimated, Figure 1.2(b). The target location is determined by using the trigonometry laws of sine and cosine [44]. Multilateration is a technique based on the measurement of the difference in distance to two or more anchors which form a hyperbolic curve [45]. The intersection of the hyperbolas, corresponding to the TDoA measurements, determines the position of the target, Figure 1.2(c).

In practice however, due to noise in radio measurements, the position lines intersect at multiple points instead of a single one. In this case, geometric approach does not provide a useful insight as to which intersection point to choose as the location of the target.

Optimization-based Techniques. If data are known to be well described by a certain statistical model, then the maximum likelihood (ML) estimator can be derived and implemented. This is because the variance of these estimators approaches asymptotically (as the signal-to-noise ratio goes high) a lower bound given by the Cramer-Rao lower bound (CRB) [46]. Typically, ML solutions are obtained as the global minimum of the non-convex objective function which is directly derived from the likelihood function of the problem. Even though a closed form ML solution is not possible because of non-linear dependence between the measurements and the unknown parameters, approximate and iterative ML techniques can be derived.

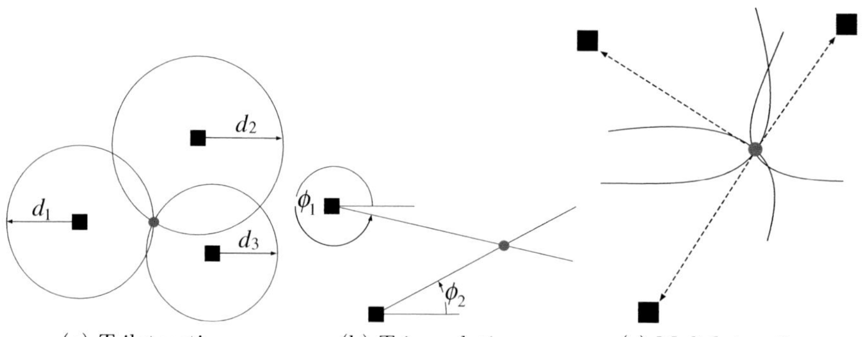

(a) Trilateration (b) Triangulation (c) Multilateration

Figure 1.2 Illustration of geometric localization techniques.

Recursive methods, such as Newton's method, combined with gradient descent method, are often used [46]. However, the objective function may have many local optima, and local search methods may easily get trapped in a local optimum. To overcome this difficulty, and possibly provide a good initial point (close to the global minimum) for the iterative algorithms, approaches such as grid search methods, linear estimators, and convex relaxation techniques have been introduced to address the ML problem [47–55]. Grid search methods solve the localization problem by forming a grid and pasing each point of the grid through the ML objective function to find the one resulting in its minimum value. These methods are time-consuming, their accuracy and computational complexity are directly proportional to the grid size and they require a huge amount of memory when the number of the unknown parameters is too large. Linear estimators are very efficient in the sense of time-consumption and computational complexity, but they are derived based on many approximations which may severely affect their performance, especially in the case when the noise is large [51]. Convex relaxation techniques overcome the difficulties in the ML problem by transforming the original non-convex and nonlinear problem into a convex one. The advantage of this approach is that the convergence to the globally optimal solution is guaranteed. However, due to application of relaxation techniques, the solution of a convex problem does not necessarily correspond to the solution of the original ML problem [56].

1.3 Outline and Contributions

This work is organized into 4 chapters. We summarize the content of each chapter, besides the current one which gives the motivation and outline of this dissertation. This work is based on ardent and dedicated research, which resulted in several publications in international journals and conferences, book chapters and patents.

In more detail, the outline and the original contributions of this work are as follows.

Chapter 2. RSS-AoA-based Target Localization: This chapter tackles the target localization problem by measurement fusion. More precisely, hybrid RSS and AoA measurements are integrated in order to enhance the estimation accuracy in comparison with *traditional* approach. By using the RSS propagation model and simple geometry, a novel objective function based on the least squares (LS) criterion is derived. For non-cooperative and cooperative WSN, the objective function is then transformed into a generalized trust region sub-problem (GTRS) and an semidefinite programming (SDP) framework, respectively. Moreover, the original non-convex target localization problem is broken down into local sub-problems, which are converted into convex problems by applying second-order cone programming (SOCP) relaxation technique. By solving this tight approximation of the original problem, each target estimates its own location, resorting only to local information from one-hop neighbors by following our described iterative procedure. Besides excellent trade-off between accuracy and computational cost, the new algorithms have a big advantage over existing approaches as their adaptation to different settings of the localization problem is straightforward.

Chapter 3. RSS-AoA-based Target Tracking: In this chapter, RSS-AoA-based target tracking problem is investigated. By applying Bayesian approach, prior knowledge given by state transition model is combined with observations, and the tracking problem via maximum *a posteriori* (MAP) criterion is formulated. Novel relationships between the unknown target location and gathered measurements are established by applying Cartesian to polar coordinates conversion, which results in efficient *linearization* of the highly non-linear observation model. By taking advantage of the *linearized* observation model and following the MAP criterion and Kalman filter (KF) recipe, novel MAP and KF algorithms are derived which efficiently solve the target tracking problem with static anchors. The target tracking problem is then extended to the where the target transmit power is not known. Also, the case where sensors' mobility is granted is studied, and a simple navigation routine is developed, which additionally betters the estimation accuracy, even for lower number of sensors.

Chapter 4. Conclusions and Future Work: This chapter concludes the work, by summarizing the main obtained results and enumerating the future lines of work.

2

RSS-AoA-based Target Localization

2.1 Chapter Summary

This chapter addresses the problem of target localization by using combined RSS and AoA measurements. It is organized into two main sections in which we study both centralized, Section 2.2, and distributed, Section 2.3, localization problems, respectively. More specifically, the remainder of the chapter is organized as follows.

Section 2.2.1 describes the SoA of the centralized RSS-AoA localization problem, and presents our contribution in that area. In Section 2.2.2, the RSS and AoA measurement models are introduced and the centralized target localization problem is formulated. Section 2.2.3 presents the development of our estimators in the case of non-cooperative localization for both known and unknown P_T. In Section 2.2.4 we describe the derivation of our centralized estimators in the case of cooperative localization for both known and unknown P_T. In Sections 2.2.5 and 2.2.6, complexity and performance analysis are presented respectively, together with the relevant results in order to compare the performance of our estimators with the SoA. Finally, Section 2.2.7 summarizes the main conclusions regarding the centralized RSS-AoA-based localization problem.

Section 2.3.1 gives an overview of the related work in the area of distributed RSS-AoA localization problem, and summarizes our contributions. In Section 2.3.2, the distributed target localization problem is formulated. Section 2.3.3 presents the development of our distributed estimators. In Section 2.3.4 we provide analysis about the computational complexity, while in Section 2.3.5 we discuss the performance of our algorithms. Finally, Section 2.3.6 summarizes the main conclusions regarding the distributed RSS-AoA-based localization problem.

2.2 Centralized RSS-AoA-based Target Localization

2.2.1 Related Work

The approaches in [49, 50, 53, 55, 57, 58] consider both non-cooperative and cooperative target localization problem, but the estimators are founded on RSS and distance measurements only. The approaches in [26–29] are based on the fusion of RSS and ToA measurements. A hybrid system that merges range and angle measurements was investigated in [30]. The authors in [30] proposed two estimators to solve the non-cooperative target localization problem in a 3-D scenario: linear LS and optimization based. The LS estimator is a relatively simple and well known estimator, while the optimization based estimator was solved by Davidon-Fletcher-Powell algorithm [59]. In [31], the authors derived an LS and an ML estimator for a hybrid scheme that combines received signal strength difference (RSSD) and AoA measurements. Non-linear constrained optimization was used to estimate the target's location from multiple RSS and AoA measurements. Both LS and ML estimators in [31] are λ-dependent, where λ is a non-negative weight assigned to regulate the contribution from RSS and AoA measurements. A selective weighted least squares (WLS) estimator for RSS/AoA localization problem was proposed in [32]. The authors determined the target location by exploiting weighted ranges from the two *nearest* anchor measurements, which were combined with the serving base station AoA measurement. In [31, 32], authors investigated the non-cooperative hybrid RSS/AoA localization problem for a 2-D scenario only. A WLS estimator for a 3-D RSSD/AoA non-cooperative localization problem when the transmit power is unknown was presented in [33]. However, the authors in [33] only investigated a small-scale WSN, with extremely low noise power. An estimator based on SDP relaxation technique for cooperative target localization problem was proposed in [60]. The authors in [60] extended their previous SDP algorithm for pure range information into a hybrid one, by adding angle information for a triplets of points. However, due to the consideration of triplets of points, the computational complexity of the SDP approach increases rather substantially with the network size.

2.2.1.1 Contribution

In this work, we investigate the target localization problem in both non-cooperative and cooperative 3-D WSN. In the case of non-cooperative WSN, we assume that all targets communicate exclusively with anchors, and a single target is located at a time. In the case of cooperative WSN, we assume that all targets communicate with any sensor within their communication range (whether it is an anchor or a target), and that all targets are located simultaneously. For both cases, a hybrid system that fuses distance and angle measurements, extracted from RSS and AoA information respectively, is employed. By using the RSS propagation model and simple geometry, we derive a novel objective function based on the LS criterion. For the case of non-cooperative WSN, based on the squared range (SR) approach we show that the derived non-convex objective function can be transformed into a GTRS framework, which can be solved exactly by a bisection procedure [28]. For the case of cooperative localization, we show that the derived objective function can be transformed into a convex function by applying SDP relaxation technique. Finally, we show that the generalization of the proposed estimators to the case where, alongside with the targets' locations, the transmit power, P_T, is also unknown, is straightforward for both non-cooperative and cooperative localization.

Thus, the main contribution of our work is threefold. First, by using RSS and AoA measurement models, we derive a novel non-convex objective function based on the LS criterion which tightly approximates the ML one for small noise. In the case of non-cooperative localization, we propose two novel estimators that significantly reduce the estimation error, compared with the state-of-the-art. Finally, in the case of cooperative localization, we present the first hybrid RSS/AoA estimators for target localization in a 3-D cooperative WSN.

2.2.2 Problem Formulation

We consider a WSN with N anchors and M targets, where the known locations of anchors are respectively denoted by $\boldsymbol{a}_1, \boldsymbol{a}_2, ..., \boldsymbol{a}_N$, and the unknown locations of targets are denoted by $\boldsymbol{x}_1, \boldsymbol{x}_2, ..., \boldsymbol{x}_M$ ($\boldsymbol{x}_i, \boldsymbol{a}_j \in \mathbb{R}^3$, $i = 1, ..., M$ and $j = 1, ..., N$). For ease of expression, let us define a vector $\boldsymbol{x} = [\boldsymbol{x}_1^T, \boldsymbol{x}_2^T, ..., \boldsymbol{x}_M^T]^T$ ($\boldsymbol{x} \in \mathbb{R}^{3M\times 1}$) as the vector of all unknown target locations, such that $\boldsymbol{x}_i = \boldsymbol{E}_i^T \boldsymbol{x}$, where $\boldsymbol{E}_i = \boldsymbol{e}_i \otimes \boldsymbol{I}_3$, and $\boldsymbol{e}_i$ is the i-th column of the identity matrix I_M. We determine these locations

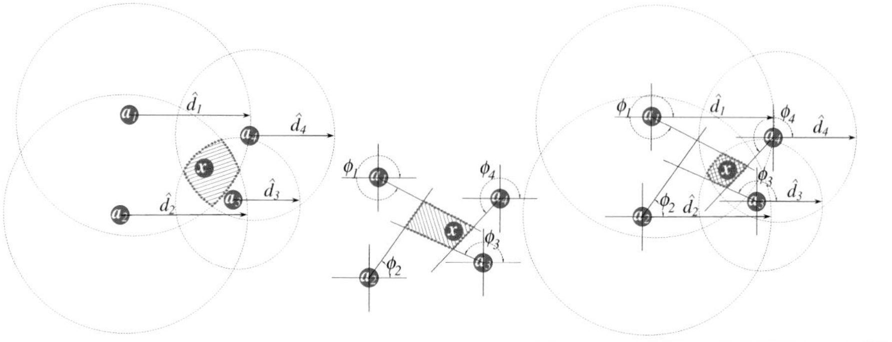

(a) Range-based Localization (b) Angle-based Localization (c) Hybrid Localization

Figure 2.1 Illustration of different localization systems in a 2-D space.

by using a hybrid system that fuses range and angle measurements. Combining two measurements of the radio signal provides more information to the user, and it is likely to enhance the estimation accuracy, as shown in Figure 2.1.

Figure 2.1 illustrates how does a (a) range-based, (b) angle-based and (c) hybrid (range and angle) system operate for the case where $M = 1$ and $N = 4$. In the range-based localization, each range measurement, $\hat{d}_i$, defines a circle as a possible location of the unknown target. Thus, a set of range measurements, $\{\hat{d}_1, \hat{d}_2, ..., \hat{d}_N\}$, defines multiple circles and the area determined by their intersection accommodates the target, Figure 2.1(a). Similarly with the angle-based localization, where each angle measurement, ϕ_i, defines a line as the set of possible locations of the unknown target, Figure 2.1(b). From Figure 2.1(c), one can see that when the two measurements of the radio signal are integrated, the set of all possible solutions (the area determined by the intersection) is significantly reduced; hence, hybrid systems are more likely to improve the estimation accuracy.

Throughout this work, it is assumed that the range measurements are obtained from the RSS information exclusively, since ranging based on RSS requires the lowest implementation costs [1]. The RSS, P_{ij}, between two sensors i and j which are within the communication range of each other (from the transmitting sensor) can be written [61, 62] as:

$$P_{ij}^{\mathcal{A}} = P_0 - 10\gamma \log_{10} \frac{\|\boldsymbol{x}_i - \boldsymbol{a}_j\|}{d_0} + n_{ij}, \text{ for } (i,j) \in \mathcal{A}, \tag{2.1a}$$

$$P_{ik}^{\mathcal{B}} = P_0 - 10\gamma \log_{10} \frac{\|\boldsymbol{x}_i - \boldsymbol{x}_k\|}{d_0} + n_{ik}, \text{ for } (i,k) \in \mathcal{B}, \tag{2.1b}$$

where P_0 (dBm) is the RSS measured at a reference distance d_0, γ is the path loss exponent, n_{ij} and n_{ik} are the log-normal shadowing terms modeled as $n_{ij} \sim \mathcal{N}(0, \sigma_{n_{ik}}^2)$, $n_{ik} \sim \mathcal{N}(0, \sigma_{n_{ik}}^2)$. Furthermore, the sets $\mathcal{A} = \{(i,j) : \|\boldsymbol{x}_i - \boldsymbol{a}_j\| \leq R, \text{for } i = 1, ..., M,\ j = 1, ..., N\}$ and $\mathcal{B} = \{(i,k) : \|\boldsymbol{x}_i - \boldsymbol{x}_k\| \leq R, \text{for } i,k = 1, ..., M, i \neq k\}$, where R is the communication range of a sensor, denote the existence of target/anchor and target/target connections, respectively.

To obtain the AoA measurements (both azimuth and elevation angles), we assume that either video cameras [63–65] or multiple antennas, or a directional antenna is implemented at anchors [30, 66, 67]. In order to make use of the AoA measurements from different sensors, the orientation information is required, which can be obtained by implementing a digital compass at each sensor [30, 66]. However, a digital compass introduces an error in the AoA measurements due to its static accuracy. For the sake of simplicity and without loss of generality, we model the angle measurement error and the orientation error as one random variable in the rest of this work.

Figure 2.2 gives an illustration of a target and anchor locations in a 3-D space. As shown in Figure 2.2, $\boldsymbol{x}_i = [x_{i1}, x_{i2}, x_{i3}]^T$ and $\boldsymbol{a}_j = [a_{j1}, a_{j2}, a_{j3}]^T$ are respectively the unknown coordinates of the i-th target and the known coordinates of the j-th anchor, while $d_{ij}^{\mathcal{A}}$, $\phi_{ij}^{\mathcal{A}}$ and $\alpha_{ij}^{\mathcal{A}}$ represent the distance, azimuth angle and elevation angle between the i-th target and the j-th anchor, respectively.

The ML estimate of the distance between two sensors can be obtained from the RSS measurement model (2.1) as follows [1]:

$$\hat{d}_{ij}^{\mathcal{A}} = d_0 10^{\frac{P_0 - P_{ij}^{\mathcal{A}}}{10\gamma}}, \text{ for } (i,j) \in \mathcal{A}, \tag{2.2a}$$

$$\hat{d}_{ik}^{\mathcal{B}} = d_0 10^{\frac{P_0 - P_{ik}^{\mathcal{B}}}{10\gamma}}, \text{ for } (i,k) \in \mathcal{B}. \tag{2.2b}$$

Applying simple geometry, azimuth and elevation angle measurements can be modeled as [30]:

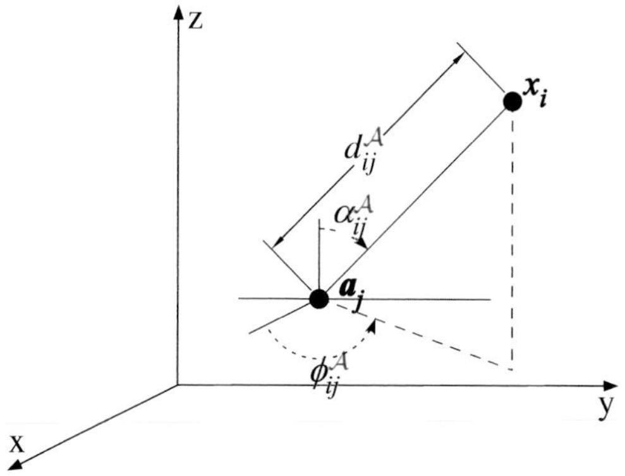

Figure 2.2 Illustration of a target and anchor locations in a 3-D space.

$$\phi_{ij}^{\mathcal{A}} = \arctan\left(\frac{x_{i2} - a_{j2}}{x_{i1} - a_{j1}}\right) + m_{ij}, \text{ for } (i,j) \in \mathcal{A}, \tag{2.3a}$$

$$\phi_{ik}^{\mathcal{B}} = \arctan\left(\frac{x_{i2} - x_{k2}}{x_{i1} - x_{k1}}\right) + m_{ik}, \text{ for } (i,k) \in \mathcal{B}, \tag{2.3b}$$

and

$$\alpha_{ij}^{\mathcal{A}} = \arccos\left(\frac{x_{i3} - a_{j3}}{\|\boldsymbol{x}_i - \boldsymbol{a}_j\|}\right) + v_{ij}, \text{ for } (i,j) \in \mathcal{A}, \tag{2.4a}$$

$$\alpha_{ik}^{\mathcal{B}} = \arccos\left(\frac{x_{i3} - x_{k3}}{\|\boldsymbol{x}_i - \boldsymbol{x}_k\|}\right) + v_{ik}, \text{ for } (i,k) \in \mathcal{B}. \tag{2.4b}$$

respectively, where m_{ij}, m_{ik} and v_{ij}, v_{ik} are respectively the measurement errors of azimuth and elevation angles, modeled as $m_{ij} \sim \mathcal{N}(0, \sigma_{m_{ij}}^2)$, $m_{ik} \sim \mathcal{N}(0, \sigma_{m_{ik}}^2)$ and $v_{ij} \sim \mathcal{N}(0, \sigma_{v_{ij}}^2)$, $v_{ik} \sim \mathcal{N}(0, \sigma_{v_{ik}}^2)$.

Given the observation vector $\boldsymbol{\theta} = [\boldsymbol{P}^T, \boldsymbol{\phi}^T, \boldsymbol{\alpha}^T]^T$ ($\boldsymbol{\theta} \in \mathbb{R}^{3(|\mathcal{A}|+|\mathcal{B}|)}$), where $\boldsymbol{P} = [P_{ij}^{\mathcal{A}}, P_{ik}^{\mathcal{B}}]^T$, $\boldsymbol{\phi} = [\phi_{ij}^{\mathcal{A}}, \phi_{ik}^{\mathcal{B}}]^T$, $\boldsymbol{\alpha} = [\alpha_{ij}^{\mathcal{A}}, \alpha_{ik}^{\mathcal{B}}]^T$, and $|\bullet|$ denotes the cardinality of a set (the number of elements in a set), the probability density function (PDF) is given as:

$$p(\boldsymbol{\theta}|\boldsymbol{x}) = \prod_{i=1}^{3(|\mathcal{A}|+|\mathcal{B}|)} \frac{1}{\sqrt{2\pi\sigma_i^2}} \exp\left\{-\frac{(\theta_i - f_i(\boldsymbol{x}))^2}{2\sigma_i^2}\right\}, \tag{2.5}$$

where

$$\boldsymbol{f}(\boldsymbol{x}) = \begin{bmatrix} \vdots \\ P_0 - 10\gamma \log_{10} \frac{\|\boldsymbol{x}_i - \boldsymbol{a}_j\|}{d_0} \\ \vdots \\ P_0 - 10\gamma \log_{10} \frac{\|\boldsymbol{x}_i - \boldsymbol{x}_k\|}{d_0} \\ \vdots \\ \arctan\left(\frac{x_{i2} - a_{j2}}{x_{i1} - a_{j1}}\right) \\ \vdots \\ \arctan\left(\frac{x_{i2} - x_{k2}}{x_{i1} - x_{k1}}\right) \\ \vdots \\ \arccos\left(\frac{x_{i3} - a_{j3}}{\|\boldsymbol{x}_i - \boldsymbol{a}_j\|}\right) \\ \vdots \\ \arccos\left(\frac{x_{i3} - x_{k3}}{\|\boldsymbol{x}_i - \boldsymbol{x}_k\|}\right) \\ \vdots \end{bmatrix}, \quad \boldsymbol{\sigma} = \begin{bmatrix} \vdots \\ \sigma_{n_{ij}} \\ \vdots \\ \sigma_{n_{ik}} \\ \vdots \\ \sigma_{m_{ij}} \\ \vdots \\ \sigma_{m_{ik}} \\ \vdots \\ \sigma_{v_{ij}} \\ \vdots \\ \sigma_{v_{ik}} \\ \vdots \end{bmatrix}.$$

The most common estimator used in practice is the ML estimator, since it has the property of being asymptotically efficient (for large enough data records) [46, 68]. The ML estimator forms its estimate as the vector $\hat{\boldsymbol{x}}$, which maximizes the PDF in (2.5); hence, the ML estimator is obtained as:

$$\hat{\boldsymbol{x}} = \arg\min_{\boldsymbol{x}} \sum_{i=1}^{3(|\mathcal{A}|+|\mathcal{B}|)} \frac{1}{\sigma_i^2} \left[\theta_i - f_i(\boldsymbol{x})\right]^2. \tag{2.6}$$

Even though the ML estimator is approximately the minimum variance unbiased estimator [46], the LS problem in (2.6) is non-convex and has no closed-form solution. In the remaining part of this work, we will show that the LS problem in (2.6) can be solved efficiently by applying certain approximations. More precisely, for non-cooperative WSN, we propose a suboptimal estimator based on the GTRS framework leading to an SR-WLS estimator, which can be solved exactly by a bisection procedure [69]. For the case of cooperative WSN, we propose a convex relaxation technique leading to an SDP estimator which can be solved efficiently by interior-point algorithms [56]. Not only that the new approaches efficiently solve the traditional RSS/AoA localization problem, but they can also be used to solve the localization problem when P_T is not known, with straightforward generalization.

2.2.2.1 Assumptions

We outline here some assumptions for the WSN (made for the sake of simplicity and without loss of generality):

(1) The network is connected and it does not change during the computation time;
(2) Measurement errors for RSS and AoA model are independent, and $\sigma_{n_{ij}} = \sigma_n$, $\sigma_{m_{ij}} = \sigma_m$ and $\sigma_{v_{ij}} = \sigma_v$, $\forall(i,j) \in \mathcal{A} \cup \mathcal{B}$;
(3) The range measurements are extracted from the RSS information exclusively and all target/target measurements are symmetric;
(4) All sensors have identical P_T;
(5) All sensors are equipped with either multiple antennas or a directional antenna, and they can measure the AoA information.

In assumption (1), we assume that the sensors are static and that there is no node/link failure during the computation period, and all sensors can convey their measurements to a central processor. Assumptions (2) and (4) are made for the sake of simplicity. Assumption (3) is made without loss of generality; it is readily seen that, if $P_{ik}^{\mathcal{B}} \neq P_{ki}^{\mathcal{B}}$, then it serves to replace $P_{ik}^{\mathcal{B}} \leftarrow (P_{ik}^{\mathcal{B}} + P_{ki}^{\mathcal{B}})/2$ and $P_{ki}^{\mathcal{B}} \leftarrow (P_{ik}^{\mathcal{B}} + P_{ki}^{\mathcal{B}})/2$ when solving the localization problem. Assumption (4) implies that P_0 and R are identical for all sensors. Finally, assumption (5) is made for the case of cooperative localization, where only some targets are able to directly connect to anchors; thus, they are forced to cooperate with other targets within their communication range.

2.2.3 Non-cooperative Localization

By non-cooperative WSN, we imply a network comprising a number of targets and anchors where each target is allowed to communicate with anchors exclusively, and a single target is localized at a time. For such a setting, we can assume that the targets are passive nodes that only emit radio signals, and that all radio measurements are collected by anchors.

In the remainder of this section, we develop a suboptimal estimator to solve the non-cooperative localization problem in (2.6), whose *exact* solution can be obtained by a bisection procedure. We then show that its generalization for the case where P_T is not known is straightforward.

2.2.3.1 Non-cooperative localization with known P_T

Note that the targets communicate with anchors exclusively in a non-cooperative network; hence, the set $\mathcal{B}$ in the path loss model (2.1) is empty. Therefore, when the noise power is sufficiently small, from (2.1a) we have:

$$\lambda_{ij}^{\mathcal{A}} \|\boldsymbol{x}_i - \boldsymbol{a}_j\| \approx d_0 \text{ for } (i,j) \in \mathcal{A}, \tag{2.7}$$

where $\lambda_{ij}^{\mathcal{A}} = 10^{\frac{P_{ij}^{\mathcal{A}} - P_0}{10\gamma}}$. Similarly, from (2.3a) and (2.4a) we respectively get:

$$\boldsymbol{c}_{ij}^T (\boldsymbol{x}_i - \boldsymbol{a}_j) \approx 0, \tag{2.8}$$

and

$$\boldsymbol{k}_{ij}^T (\boldsymbol{x}_i - \boldsymbol{a}_j) \approx \|\boldsymbol{x}_i - \boldsymbol{a}_j\| \cos(\alpha_{ij}^{\mathcal{A}}), \tag{2.9}$$

where $\boldsymbol{c}_{ij} = [-\sin(\phi_{ij}^{\mathcal{A}}), \cos(\phi_{ij}^{\mathcal{A}}), 0]^T$ and $\boldsymbol{k}_{ij} = [0, 0, 1]^T$.

Next, we can rewrite (2.7) as:

$$\lambda_{ij}^{\mathcal{A}^2} \|\boldsymbol{x}_i - \boldsymbol{a}_j\|^2 \approx d_0^2. \tag{2.10}$$

Introduce weights, $\boldsymbol{w} = [\sqrt{w_{ij}}]$, where each w_{ij} is defined as

$$w_{ij} = 1 - \frac{\hat{d}_{ij}^{\mathcal{A}}}{\sum_{(i,j)\in\mathcal{A}} \hat{d}_{ij}^{\mathcal{A}}},$$

such that more importance is given to nearby links. The reason for defining the weights in this manner is because both RSS and AoA short-range measurements are trusted more than long ones. The RSS measurements have relatively constant standard deviation with distance [1]. This implies that multiplicative factors of RSS measurements are constant with range. For example, for a multiplicative factor of 1.5, at a range of 1 m, the measured range would be 1.5 m, and at an actual range of 10 m, the measured range would be 15 m, a factor of 10 times greater [1]. In the case of AoA measurements, the reason is more intuitive, and we call the reader's attention to Figure 2.3.

In Figure 2.3, an azimuth angle measurement made between an anchor and two targets located along the same line, but with different distances from the anchor is illustrated. The true and the measured

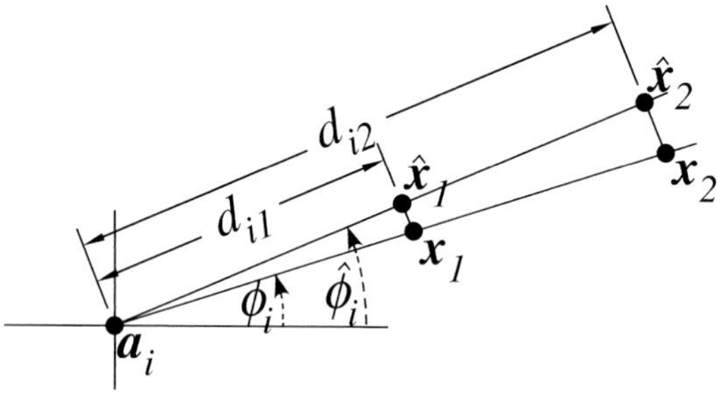

Figure 2.3 Illustration of azimuth angle measurements: short-range versus long-range.

azimuth angles between the anchor and the targets are denoted by ϕ_i and $\hat{\phi}_i$, respectively. Our goal is to determine the locations of the two targets. Based on the available information, the location estimates of the two targets are at points $\hat{\boldsymbol{x}}_1$ and $\hat{\boldsymbol{x}}_2$. However, from Figure 2.3, we can see that the estimated location of the target physically closer to the anchor ($\hat{\boldsymbol{x}}_1$) is much closer to its true location than the one further away. In other words, for a given angle, the more two sensors are physically further apart the greater the set of all possible solutions will be (more likely to impair the localization accuracy).

Replace $\|\boldsymbol{x}_i - \boldsymbol{a}_j\|$ in (2.9) with $\hat{d}_{ij}^{\mathcal{A}}$ described in (2.2a), to obtain the following WLS problem according to (2.10), (2.8) and (2.9) as:

$$\hat{\boldsymbol{x}}_i = \arg\min_{\boldsymbol{x}_i} \sum_{(i,j):(i,j)\in\mathcal{A}} w_{ij}\left(\lambda_{ij}^{\mathcal{A}^2}\|\boldsymbol{x}_i-\boldsymbol{a}_j\|^2-d_0^2\right)^2+\sum_{(i,j):(i,j)\in\mathcal{A}} w_{ij}\left(\boldsymbol{c}_{ij}^T(\boldsymbol{x}_i-\boldsymbol{a}_j)\right)^2$$
$$+\sum_{(i,j):(i,j)\in\mathcal{A}} w_{ij}\left(\boldsymbol{k}_{ij}^T(\boldsymbol{x}_i-\boldsymbol{a}_j)-\hat{d}_{ij}^{\mathcal{A}}\cos(\alpha_{ij}^{\mathcal{A}})\right)^2. \tag{2.11}$$

The above WLS estimator is non-convex and has no closed-form solution. However, we can express (2.11) as a quadratic programming problem whose *global* solution can be computed efficiently [69]. Using the substitution $\boldsymbol{y}_i = [\boldsymbol{x}_i^T, \|\boldsymbol{x}_i\|^2]^T$, the problem in (2.11) can be rewritten as:

$$\underset{\boldsymbol{y}_i}{\text{minimize}} \; \|\boldsymbol{W}(\boldsymbol{A}\boldsymbol{y}_i - \boldsymbol{b})\|^2$$

subject to

$$\boldsymbol{y}_i^T\boldsymbol{D}\boldsymbol{y}_i + 2\boldsymbol{l}^T\boldsymbol{y}_i = 0, \tag{2.12}$$

where $\boldsymbol{W} = \boldsymbol{I}_3 \otimes \text{diag}(\boldsymbol{w})$,

$$\boldsymbol{A} = \begin{bmatrix} \vdots & \vdots \\ -2{\lambda_{ij}^{\mathcal{A}}}^2 \boldsymbol{a}_j^T & {\lambda_{ij}^{\mathcal{A}}}^2 \\ \vdots & \vdots \\ \boldsymbol{c}_{ij}^T & 0 \\ \vdots & \vdots \\ \boldsymbol{k}_{ij}^T & 0 \\ \vdots & \vdots \end{bmatrix}, \boldsymbol{b} = \begin{bmatrix} \vdots \\ d_0^2 - {\lambda_{ij}^{\mathcal{A}}}^2 \|\boldsymbol{a}_j\|^2 \\ \vdots \\ \boldsymbol{c}_{ij}^T \boldsymbol{a}_j \\ \vdots \\ \boldsymbol{k}_{ij}^T \boldsymbol{a}_j + \hat{d}_{ij}^{\mathcal{A}} \cos(\alpha_{ij}^{\mathcal{A}}) \\ \vdots \end{bmatrix},$$

$$\boldsymbol{D} = \begin{bmatrix} \boldsymbol{I}_3 & \boldsymbol{0}_{3\times 1} \\ \boldsymbol{0}_{1\times 3} & 0 \end{bmatrix}, \boldsymbol{l} = \begin{bmatrix} \boldsymbol{0}_{3\times 1} \\ -1/2 \end{bmatrix},$$

i.e., $\boldsymbol{A} \in \mathbb{R}^{3|\mathcal{A}|\times 4}$, $\boldsymbol{b} \in \mathbb{R}^{3|\mathcal{A}|\times 1}$, and $\boldsymbol{W} \in \mathbb{R}^{3|\mathcal{A}|\times 3|\mathcal{A}|}$.

The objective function and the constraint in (2.12) are both quadratic. This type of problem is known as GTRS [69, 70], and it can be solved exactly by a bisection procedure [69]. We denote (2.12) as "SR-WLS1" in the remaining text.

2.2.3.2 Non-cooperative localization with unknown P_T

To maintain low implementation costs, testing and calibration are not the priority in practice. Thus, sensors' transmit powers are often not calibrated, *i.e.*, not known. Not knowing P_T in the RSS measurement model corresponds to not knowing P_0 in the RSS model (2.1); see [16, 38] and the references therein.

The generalization of the proposed estimators for known P_0 is straightforward for the case where P_0 is not known. Notice that Equation (2.7) can be rewritten as:

$$\beta_{ij}^{\mathcal{A}} \|\boldsymbol{x}_i - \boldsymbol{a}_j\| \approx \eta d_0, \text{ for } (i,j) \in \mathcal{A}, \tag{2.13}$$

where $\beta_{ij}^{\mathcal{A}} = 10^{\frac{P_{ij}^{\mathcal{A}}}{10\gamma}}$, and $\eta = 10^{\frac{P_0}{10\gamma}}$ is an unknown parameter that needs to be estimated.

Substitute $\|\boldsymbol{x}_i - \boldsymbol{a}_j\|$ with $\hat{d}_{ij}^{\mathcal{A}}$ in (2.9). Then, we can rewrite (2.9) as:

$$\beta_{ij}^{\mathcal{A}} \boldsymbol{k}_{ij}^T (\boldsymbol{x}_i - \boldsymbol{a}_j) \approx \eta d_0 \cos(\alpha_{ij}^{\mathcal{A}}). \tag{2.14}$$

In order to assign more importance to nearby links, introduce weights $\widetilde{\boldsymbol{w}} = [\sqrt{\widetilde{w}_{ij}}]$, where

$$\widetilde{w}_{ij} = \frac{P_{ij}^{\mathcal{A}}}{\sum_{(i,j)\in\mathcal{A}} P_{ij}^{\mathcal{A}}}.$$

By squaring (2.13), we can obtain the following WLS problem, according to (2.13), (2.8) and (2.14) as:

$$(\hat{\boldsymbol{x}}_i, \hat{\eta}) = \arg\min_{\boldsymbol{x}_i,\eta} \sum_{(i,j):(i,j)\in\mathcal{A}} \widetilde{w}_{ij} \left({\beta_{ij}^{\mathcal{A}}}^2 \|\boldsymbol{x}_i - \boldsymbol{a}_j\|^2 - \eta^2 d_0^2\right)^2$$
$$+ \sum_{(i,j):(i,j)\in\mathcal{A}} \widetilde{w}_{ij}\left(\boldsymbol{c}_{ij}^T(\boldsymbol{x}_i - \boldsymbol{a}_j)\right)^2 + \sum_{(i,j):(i,j)\in\mathcal{A}} \widetilde{w}_{ij}\left(\beta_{ij}^{\mathcal{A}}\boldsymbol{k}_{ij}^T(\boldsymbol{x}_i - \boldsymbol{a}_j) - \eta d_0 \cos(\alpha_{ij}^{\mathcal{A}})\right)^2.$$

Using the substitution $\widetilde{\boldsymbol{y}}_i = [\boldsymbol{x}_i^T, \|\boldsymbol{x}_i\|^2, \eta, \eta^2]^T$, we can rewrite (2.15) as a GTRS:

$$\underset{\widetilde{\boldsymbol{y}}_i}{\text{minimize}} \ \|\widetilde{\boldsymbol{W}}(\widetilde{\boldsymbol{A}}\widetilde{\boldsymbol{y}}_i - \widetilde{\boldsymbol{b}})\|^2$$

subject to

$$\widetilde{\boldsymbol{y}}_i^T \widetilde{\boldsymbol{D}} \widetilde{\boldsymbol{y}}_i + 2\widetilde{\boldsymbol{l}}^T \widetilde{\boldsymbol{y}}_i = 0, \tag{2.15}$$

where $\widetilde{\boldsymbol{W}} = \boldsymbol{I}_3 \otimes \text{diag}(\widetilde{\boldsymbol{w}})$, $\widetilde{\boldsymbol{D}} = \text{diag}([1,1,1,0,1,0])$, and

$$\widetilde{\boldsymbol{A}} = \begin{bmatrix} \vdots & \vdots & \vdots & \vdots \\ -2{\beta_{ij}^{\mathcal{A}}}^2 \boldsymbol{a}_j^T & {\beta_{ij}^{\mathcal{A}}}^2 & 0 & -d_0 \\ \vdots & \vdots & \vdots & \vdots \\ \boldsymbol{c}_{ij}^T & 0 & 0 & 0 \\ \vdots & \vdots & \vdots & \vdots \\ \beta_{ij}^{\mathcal{A}}\boldsymbol{k}_{ij}^T & 0 & -d_0\cos(\alpha_{ij}^{\mathcal{A}}) & 0 \\ \vdots & \vdots & \vdots & \vdots \end{bmatrix}, \widetilde{\boldsymbol{b}} = \begin{bmatrix} \vdots \\ -{\beta_{ij}^{\mathcal{A}}}^2\|\boldsymbol{a}_j\|^2 \\ \vdots \\ \boldsymbol{c}_{ij}^T\boldsymbol{a}_j \\ \vdots \\ \beta_{ij}^{\mathcal{A}}\boldsymbol{k}_{ij}^T\boldsymbol{a}_j \\ \vdots \end{bmatrix},$$

$$\widetilde{\boldsymbol{l}} = \left[\boldsymbol{0}_{1\times 3}, -\frac{1}{2}, 0, -\frac{1}{2}\right]^T,$$

i.e., $\widetilde{\boldsymbol{A}} \in \mathbb{R}^{3|\mathcal{A}|\times 6}$, $\widetilde{\boldsymbol{b}} \in \mathbb{R}^{3|\mathcal{A}|\times 1}$, and $\widetilde{\boldsymbol{W}} \in \mathbb{R}^{3|\mathcal{A}|\times 3|\mathcal{A}|}$.

Even though the approach in (2.15) efficiently solves (2.6) for unknown P_0, we can further improve its performance. To do so, we will first solve (2.15) to obtain the location estimate, and use this estimate to find the ML estimate of P_0, $\widehat{P}_0$. Then, we will take advantage of $\widehat{P}_0$

to solve another WLS problem as if P_0 is known. Hence, the proposed procedure for solving (2.6) when P_0 is not known is summarized below:

1. Solve (2.15) to obtain the initial estimate of $\boldsymbol{x}_i$, $\hat{\boldsymbol{x}}_i'$;
2. Use $\hat{\boldsymbol{x}}_i'$ to compute the ML estimate of P_0, $\widehat{P}_0$ as:

$$\widehat{P}_0 = \frac{\sum_{(i,j)\in\mathcal{A}} \left(P_{ij}^{\mathcal{A}} + 10\gamma \log_{10} \frac{\|\hat{\boldsymbol{x}}_i' - \boldsymbol{a}_j\|}{d_0} \right)}{|\mathcal{A}|};$$

3. Exploit $\widehat{P}_0$ to calculate $\hat{\lambda}_{ij}^{\mathcal{A}} = 10^{\frac{P_{ij}^{\mathcal{A}} - \widehat{P}_0}{10\gamma}}$, and use this estimated value to solve the SR-WLS in (2.12).

The main reason for applying this simple procedure is that we observed in our simulations that after solving (2.15) an excellent ML estimation of P_0, $\widehat{P}_0$, is obtained, very close to the true value of P_0. This motivated us to take advantage of this estimated value to solve another WLS problem (2.12), as if P_0 is known. We denote the above three-step procedure as "SR-WLS2" in the remaining text.

2.2.4 Cooperative Localization

By cooperative WSN, we imply a network consisting of a number of targets and anchors where a target can communicate with any sensor within its communication range, and all targets are localized simultaneously. A kind of node cooperation is required in networks with modest energy capabilities, where communication ranges are limited (in order to prolong the sensors' battery lives) and only some targets can communicate directly with the anchor nodes.

Throughout this section, we develop a convex estimator by using appropriate relaxation technique leading to an SDP estimator for 3-D localization. Moreover, we show that the generalization of the proposed estimator to the case of unknown P_0 is straightforward.

For sufficiently small noise, (2.1), (2.3) and (2.4) can be rewritten as:

$$\lambda_{ij}^{\mathcal{A}^2} \|\boldsymbol{x}_i - \boldsymbol{a}_j\|^2 \approx d_0^2, \text{ for } (i,j) \in \mathcal{A}, \tag{2.16a}$$

$$\lambda_{ik}^{\mathcal{B}^2} \|\boldsymbol{x}_i - \boldsymbol{x}_k\|^2 \approx d_0^2, \text{ for } (i,k) \in \mathcal{B}, \tag{2.16b}$$

$$\boldsymbol{c}_{ij}^T(\boldsymbol{x}_i - \boldsymbol{a}_j) \approx 0, \text{ for } (i,j) \in \mathcal{A}, \tag{2.17a}$$
$$\boldsymbol{c}_{ik}^T(\boldsymbol{x}_i - \boldsymbol{x}_k) \approx 0, \text{ for } (i,k) \in \mathcal{B}, \tag{2.17b}$$

and

$$\boldsymbol{k}_{ij}^T(\boldsymbol{x}_i - \boldsymbol{a}_j)(\boldsymbol{x}_i - \boldsymbol{a}_j)^T\boldsymbol{k}_{ij} \approx \|\boldsymbol{x}_i - \boldsymbol{a}_j\|^2\cos^2(\alpha_{ij}^{\mathcal{A}}), \text{ for } (i,j) \in \mathcal{A}, \tag{2.18a}$$
$$\boldsymbol{k}_{ik}^T(\boldsymbol{x}_i - \boldsymbol{x}_k)(\boldsymbol{x}_i - \boldsymbol{x}_k)^T\boldsymbol{k}_{ik} \approx \|\boldsymbol{x}_i - \boldsymbol{x}_k\|^2\cos^2(\alpha_{ik}^{\mathcal{B}}), \text{ for } (i,k) \in \mathcal{B}, \tag{2.18b}$$

where $\lambda_{ik}^{\mathcal{B}} = 10^{\frac{P_{ik}^{\mathcal{B}} - P_0}{10\gamma}}$, $\boldsymbol{c}_{ik} = [-\sin(\phi_{ik}^{\mathcal{B}}), \cos(\phi_{ik}^{\mathcal{B}}), 0]^T$, and $\boldsymbol{k}_{ik} = [0,0,1]^T$.

Following the LS principle, from (2.16), (2.17) and (2.18) we obtain the target location estimates, $\hat{\boldsymbol{x}}$, by minimizing the objective function:

$$\begin{aligned}\hat{\boldsymbol{x}} = \arg\min_{\boldsymbol{x}} & \sum_{(i,j):(i,j)\in\mathcal{A}} \left(\lambda_{ij}^{\mathcal{A}^2}\|\boldsymbol{x}_i - \boldsymbol{a}_j\|^2 - d_0^2\right)^2 + \sum_{(i,j):(i,j)\in\mathcal{A}} \left(\boldsymbol{c}_{ij}^T(\boldsymbol{x}_i - \boldsymbol{a}_j)\right)^2 \\ & + \sum_{(i,j):(i,j)\in\mathcal{A}} \left(\boldsymbol{k}_{ij}^T(\boldsymbol{x}_i - \boldsymbol{a}_j)(\boldsymbol{x}_i - \boldsymbol{a}_j)^T\boldsymbol{k}_{ij} - \|\boldsymbol{x}_i - \boldsymbol{a}_j\|^2\cos^2(\alpha_{ij}^{\mathcal{A}})\right)^2 \\ & + \sum_{(i,k):(i,k)\in\mathcal{B}} \left(\lambda_{ik}^{\mathcal{B}^2}\|\boldsymbol{x}_i - \boldsymbol{x}_k\|^2 - d_0^2\right)^2 + \sum_{(i,k):(i,k)\in\mathcal{B}} \left(\boldsymbol{c}_{ik}^T(\boldsymbol{x}_i - \boldsymbol{x}_k)\right)^2 \\ & + \sum_{(i,k):(i,k)\in\mathcal{B}} \left(\boldsymbol{k}_{ik}^T(\boldsymbol{x}_i - \boldsymbol{x}_k)(\boldsymbol{x}_i - \boldsymbol{x}_k)^T\boldsymbol{k}_{ik} - \|\boldsymbol{x}_i - \boldsymbol{x}_k\|^2\cos^2(\alpha_{ik}^{\mathcal{B}})\right)^2.\end{aligned} \tag{2.19}$$

Although the optimization problem in (2.19) is non-convex and has no closed-form solution, we will show in the following text that it can be converted into an SDP problem.

2.2.4.1 Cooperative localization with known P_T

A common approach in the literature (when dealing with cooperative localization) is to stack all unknowns in one big matrix variable $\boldsymbol{Y} = [\boldsymbol{x}_1, ..., \boldsymbol{x}_M]$ ($\boldsymbol{Y} \in \mathbb{R}^{3\times M}$) [49–55]. However, this approach cannot be applied to solve (2.19) because of the vector outer product that appears in two sums with respect to the elevation angle. Instead, we assemble the unknowns in a vector, which allows us to cope effortlessly with the outer product.

Introduce auxiliary variable $\boldsymbol{X} = \boldsymbol{x}\boldsymbol{x}^T$ ($\boldsymbol{X} \in \mathbb{R}^{3M\times 3M}$). Moreover, introduce an auxiliary vector $\boldsymbol{z} = [z_{ij}^{\mathcal{A}}, g_{ij}^{\mathcal{A}}, p_{ij}^{\mathcal{A}}, z_{ik}^{\mathcal{B}}, g_{ik}^{\mathcal{B}}, p_{ik}^{\mathcal{B}}]^T$ ($\boldsymbol{z} \in \mathbb{R}^{3(|\mathcal{A}|+|\mathcal{B}|)\times 1}$). Then, the problem in (2.19) can be rewritten as:

$$\underset{\boldsymbol{x},\boldsymbol{X},\boldsymbol{z}}{\text{minimize}} \ \|\boldsymbol{z}\|^2$$

subject to

$$z_{ij}^{\mathcal{A}} = \lambda_{ij}^{\mathcal{A}2}\left(\text{tr}\left(\boldsymbol{E}_i^T\boldsymbol{X}\boldsymbol{E}_i\right) - 2\boldsymbol{a}_j^T\boldsymbol{E}_i^T\boldsymbol{x} + \|\boldsymbol{a}_j\|^2\right) - d_0^2, \tag{2.20a}$$

$$g_{ij}^{\mathcal{A}} = \boldsymbol{c}_{ij}^T\left(\boldsymbol{E}_i^T\boldsymbol{x} - \boldsymbol{a}_j\right), \ \text{for } (i,j) \in \mathcal{A}, \tag{2.20b}$$

$$p_{ij}^{\mathcal{A}} = \boldsymbol{k}_{ij}^T\left(\boldsymbol{E}_i^T\boldsymbol{X}\boldsymbol{E}_i - 2\boldsymbol{E}_i^T\boldsymbol{x}\boldsymbol{a}_j^T + \boldsymbol{a}_j\boldsymbol{a}_j^T\right)\boldsymbol{k}_{ij}$$
$$-\left(\text{tr}\left(\boldsymbol{E}_i^T\boldsymbol{X}\boldsymbol{E}_i\right) - 2\boldsymbol{a}_j^T\boldsymbol{E}_i^T\boldsymbol{x} + \|\boldsymbol{a}_j\|^2\right)\cos^2(\alpha_{ij}^{\mathcal{A}}), \tag{2.20c}$$

$$z_{ik}^{\mathcal{B}} = \lambda_{ik}^{\mathcal{B}\,2}\left(\text{tr}\left(\boldsymbol{E}_i^T\boldsymbol{X}\boldsymbol{E}_i\right) - 2\,\text{tr}\left(\boldsymbol{E}_i^T\boldsymbol{X}\boldsymbol{E}_k\right) + \text{tr}\left(\boldsymbol{E}_{\boldsymbol{k}}^T\boldsymbol{X}\boldsymbol{E}_{\boldsymbol{k}}\right)\right) - d_0^2, \tag{2.20d}$$

$$g_{ik}^{\mathcal{B}} = \boldsymbol{c}_{ik}^T\left(\boldsymbol{E}_i^T\boldsymbol{x} - \boldsymbol{E}_k^T\boldsymbol{x}\right), \ \text{for } (i,k) \in \mathcal{B} \tag{2.20e}$$

$$p_{ik}^{\mathcal{B}} = \boldsymbol{k}_{ik}^T\left(\boldsymbol{E}_i^T\boldsymbol{X}\boldsymbol{E}_i - 2\boldsymbol{E}_i^T\boldsymbol{X}\boldsymbol{E}_k + \boldsymbol{E}_k^T\boldsymbol{X}\boldsymbol{E}_k\right)\boldsymbol{k}_{ik}$$
$$-\left(\text{tr}\left(\boldsymbol{E}_i^T\boldsymbol{X}\boldsymbol{E}_i\right) - 2\,\text{tr}\left(\boldsymbol{E}_i^T\boldsymbol{X}\boldsymbol{E}_k\right) + \text{tr}\left(\boldsymbol{E}_k^T\boldsymbol{X}\boldsymbol{E}_k\right)\right)\cos^2(\alpha_{ik}^{\mathcal{B}}), \tag{2.20f}$$

$$\boldsymbol{X} = \boldsymbol{x}\boldsymbol{x}^T. \tag{2.20g}$$

Defining an epigraph variable, t, together with the semidefinite and second-order cone relaxations of the form $\boldsymbol{X} \succeq \boldsymbol{x}\boldsymbol{x}^T$ and $\|\boldsymbol{z}\|^2 \leq t$ respectively, the following convex epigraph form is obtained from the above problem:

$$\underset{\boldsymbol{x},\boldsymbol{X},\boldsymbol{z},t}{\text{minimize}} \ t$$

subject to (2.20a)–(2.20f),

$$\left\|\begin{bmatrix} 2\boldsymbol{z} \\ t-1 \end{bmatrix}\right\| \leq t+1, \ \begin{bmatrix} \boldsymbol{X} & \boldsymbol{x} \\ \boldsymbol{x}^T & 1 \end{bmatrix} \succeq \boldsymbol{0}_{3M+1}. \tag{2.21}$$

The above problem is an SDP (more precisely, it is a mixed SDP/SOCP), which can be readily solved by `CVX` [71]. It is worth mentioning that, if rank$(\boldsymbol{X}) = 1$, then the relaxed constraint $\boldsymbol{X} \succeq \boldsymbol{x}\boldsymbol{x}^T$ is satisfied as an equality [56]. Noteworthily, we applied the Schur complement to rewrite $\boldsymbol{X} \succeq \boldsymbol{x}\boldsymbol{x}^T$ into a semidefinite cone constraint form. In the following text, we will denote (2.21) as "SDP1".

2.2.4.2 Cooperative localization with unknown P_T

The generalization of the proposed SDP estimator for known P_0 is straightforward for the case where P_0 is not known. From (2.16), we have that:

$$\beta_{ij}^{\mathcal{A}^2} \|\boldsymbol{x}_i - \boldsymbol{a}_j\|^2 \approx \rho d_0^2, \text{ for } (i,j) \in \mathcal{A}, \tag{2.22a}$$

$$\beta_{ik}^{\mathcal{B}^2} \|\boldsymbol{x}_i - \boldsymbol{x}_k\|^2 \approx \rho d_0^2, \text{ for } (i,k) \in \mathcal{B}, \tag{2.22b}$$

where $\beta_{ik} = 10^{\frac{P_{ik}^{\mathcal{B}}}{10\gamma}}$ for $(i,k) \in \mathcal{B}$ and $\rho = 10^{\frac{P_0}{5\gamma}}$. Therefore, according to (2.22), (2.17) and (2.18), the target location estimates are obtained by minimizing the following LS problem:

$$\begin{aligned}(\hat{\boldsymbol{x}}, \hat{\rho}) = \underset{\boldsymbol{x},\rho}{\arg\min} & \sum_{(i,j):(i,j)\in\mathcal{A}} \left(\beta_{ij}^{\mathcal{A}^2} \|\boldsymbol{x}_i - \boldsymbol{a}_j\|^2 - \rho d_0^2\right)^2 + \sum_{(i,j):(i,j)\in\mathcal{A}} \left(\boldsymbol{c}_{ij}^T(\boldsymbol{x}_i - \boldsymbol{a}_j)\right)^2 \\ & + \sum_{(i,j):(i,j)\in\mathcal{A}} \left(\boldsymbol{k}_{ij}^T (\boldsymbol{x}_i - \boldsymbol{a}_j)(\boldsymbol{x}_i - \boldsymbol{a}_j)^T \boldsymbol{k}_{ij} - \|\boldsymbol{x}_i - \boldsymbol{a}_j\|^2 \cos^2(\alpha_{ij}^{\mathcal{A}})\right)^2 \\ & + \sum_{(i,k):(i,k)\in\mathcal{B}} \left(\beta_{ik}^{\mathcal{B}^2} \|\boldsymbol{x}_i - \boldsymbol{x}_k\|^2 - \rho d_0^2\right)^2 + \sum_{(i,k):(i,k)\in\mathcal{B}} \left(\boldsymbol{c}_{ik}^T(\boldsymbol{x}_i - \boldsymbol{x}_k)\right)^2 \\ & + \sum_{(i,k):(i,k)\in\mathcal{B}} \left(\boldsymbol{k}_{ik}^T (\boldsymbol{x}_i - \boldsymbol{x}_k)(\boldsymbol{x}_i - \boldsymbol{x}_k)^T \boldsymbol{k}_{ik} - \|\boldsymbol{x}_i - \boldsymbol{x}_k\|^2 \cos^2(\alpha_{ik}^{\mathcal{B}})\right)^2.\end{aligned} \tag{2.23}$$

By following similar steps as described in the previous section, we obtain the SDP estimator defined below:

$$\underset{\boldsymbol{x},\rho,\boldsymbol{X},\boldsymbol{z},t}{\text{minimize}} \; t$$

subject to

$$\begin{aligned} z_{ij}^{\mathcal{A}} &= \beta_{ij}^{\mathcal{A}^2}\left(\text{tr}\left(\boldsymbol{E}_i^T\boldsymbol{X}\boldsymbol{E}_i\right) - 2\boldsymbol{a}_j^T\boldsymbol{E}_i^T\boldsymbol{x} + \|\boldsymbol{a}_j\|^2\right) - \rho d_0^2, \\ g_{ij}^{\mathcal{A}} &= \boldsymbol{c}_{ij}^T\left(\boldsymbol{E}_i^T\boldsymbol{x} - \boldsymbol{a}_j\right), \text{ for } (i,j) \in \mathcal{A}, \\ p_{ij}^{\mathcal{A}} &= \boldsymbol{k}_{ij}^T\left(\boldsymbol{E}_i^T\boldsymbol{X}\boldsymbol{E}_i - 2\boldsymbol{E}_i^T\boldsymbol{x}\boldsymbol{a}_j^T + \boldsymbol{a}_j\boldsymbol{a}_j^T\right)\boldsymbol{k}_{ij} \\ &\quad - \left(\text{tr}\left(\boldsymbol{E}_i^T\boldsymbol{X}\boldsymbol{E}_i\right) - 2\boldsymbol{a}_j^T\boldsymbol{E}_i^T\boldsymbol{x} + \|\boldsymbol{a}_j\|^2\right)\cos^2(\alpha_{ij}^{\mathcal{A}}), \\ z_{ik}^{\mathcal{B}} &= \beta_{ik}^{\mathcal{B}^2}\left(\text{tr}\left(\boldsymbol{E}_i^T\boldsymbol{X}\boldsymbol{E}_i\right) - 2\,\text{tr}\left(\boldsymbol{E}_i^T\boldsymbol{X}\boldsymbol{E}_k\right) + \text{tr}\left(\boldsymbol{E}_k^T\boldsymbol{X}\boldsymbol{E}_k\right)\right) - \rho d_0^2, \\ g_{ik}^{\mathcal{B}} &= \boldsymbol{c}_{ik}^T\left(\boldsymbol{E}_i^T\boldsymbol{x} - \boldsymbol{E}_k^T\boldsymbol{x}\right), \text{ for } (i,k) \in \mathcal{B} \\ p_{ik}^{\mathcal{B}} &= \boldsymbol{k}_{ik}^T\left(\boldsymbol{E}_i^T\boldsymbol{X}\boldsymbol{E}_i - 2\boldsymbol{E}_i^T\boldsymbol{X}\boldsymbol{E}_k + \boldsymbol{E}_k^T\boldsymbol{X}\boldsymbol{E}_k\right)\boldsymbol{k}_{ik} \\ &\quad - \left(\text{tr}\left(\boldsymbol{E}_i^T\boldsymbol{X}\boldsymbol{E}_i\right) - 2\,\text{tr}\left(\boldsymbol{E}_i^T\boldsymbol{X}\boldsymbol{E}_k\right) + \text{tr}\left(\boldsymbol{E}_k^T\boldsymbol{X}\boldsymbol{E}_k\right)\right)\cos^2(\alpha_{ik}^{\mathcal{B}}), \end{aligned}$$

$$\left\|\begin{bmatrix} 2\boldsymbol{z} \\ t-1 \end{bmatrix}\right\| \leq t+1, \ \begin{bmatrix} \boldsymbol{X} & \boldsymbol{x} \\ \boldsymbol{x}^T & 1 \end{bmatrix} \succeq \mathbf{0}_{3M+1}. \tag{2.24}$$

Although the above SDP estimator efficiently solves the target localization problem for the case of unknown P_0 in a cooperative WSN, we propose the following three-step procedure to further enhance the estimation accuracy:

1. Solve (2.24) to obtain the initial estimate of all target locations $\boldsymbol{x}$, $\hat{\boldsymbol{x}}'$;
2. Use $\hat{\boldsymbol{x}}'$ to compute the ML estimate of P_0, $\widehat{P}_0$ as:

$$\widehat{P}_0 = \frac{\sum_{(i,j):(i,j)\in\mathcal{A}} \left(P_{ij}^{\mathcal{A}} + 10\gamma\log_{10}\frac{\|\boldsymbol{E}_i^T\hat{\boldsymbol{x}}' - \boldsymbol{a}_j\|}{d_0}\right)}{|\mathcal{A}| + |\mathcal{B}|} + \frac{\sum_{(i,k):(i,k)\in\mathcal{B}} \left(P_{ik}^{\mathcal{B}} + 10\gamma\log_{10}\frac{\|\boldsymbol{E}_i^T\hat{\boldsymbol{x}}' - \boldsymbol{E}_k^T\hat{\boldsymbol{x}}'\|}{d_0}\right)}{|\mathcal{A}| + |\mathcal{B}|};$$

3. Exploit $\widehat{P}_0$ to calculate $\hat{\lambda}_{ij}^{\mathcal{A}} = 10^{\frac{P_{ij}^{\mathcal{A}} - \widehat{P}_0}{10\gamma}}$ $\forall(i,j) \in \mathcal{A}$ and $\hat{\lambda}_{ik}^{\mathcal{B}} = 10^{\frac{P_{ik}^{\mathcal{B}} - \widehat{P}_0}{10\gamma}}$ $\forall(i,k) \in \mathcal{B}$, and use these estimated values to solve the SDP in (2.21) as if P_0 is known.

We refer to the above three-step procedure as "SDP2" in the following text.

2.2.5 Complexity Analysis

The trade-off between the estimation accuracy and the computational complexity is one of the most important features of any algorithm since it defines its applicability potential. This is the reason why, apart from the performance, we also want to analyse the computational complexity of the considered approaches.

The following formula for computing the worst case computational complexity of a mixed SDP/SOCP [72] is used to analyse the complexities of the considered algorithms in this section. For completeness, we reproduce the formula below

$$\mathcal{O}\left(\sqrt{L}\left(m\sum_{i=1}^{N_{sd}} n_i^{sd^3} + m^2\sum_{i=1}^{N_{sd}} n_i^{sd^2} + m^2\sum_{i=1}^{N_{soc}} n_i^{soc} + \sum_{i=1}^{N_{soc}} n_i^{soc^2} + m^3\right)\right),$$

where L is the iteration complexity of the algorithm, m is the number of equality constraints, n_i^{sd} and n_i^{soc} are respectively the dimensions of the i-th semidefinite cone (SDC) and the i-th second-order cone (SOC), and N_i^{sd} and N_i^{soc} are the number of SDC and SOC constraints, respectively. The above formula corresponds to the formula for computing the complexity of an SDP for the case when we have no SOCCs (in which case L is the dimension of the SDC given as a result of accumulating all SDC), and vice versa (in which case L is the total number of SOC constraints) [72].

Since we are interested in analysing the worst case asymptotic computational complexity, we present only the dominating elements, which are expressed as a function of N and M. Therefore, we assume that the network is fully connected[1], *i.e.*, the total number of connections in the network is $C = |\mathcal{A}| + |\mathcal{B}|$, where $|\mathcal{A}| = MN$ and $|\mathcal{B}| = \frac{M(M-1)}{2}$. In the case of non-cooperative localization, each target is located at a time; hence, we can presume that $M = 1$ in this case.

Assuming that $K_{\max}$ is the maximum number of steps in the bisection procedure used to solve (2.12) and (2.15), Table 2.1 provides an overview of the considered algorithms together with their worst case computational complexities.

Table 2.1 reveals that the computational complexity of the considered approaches depends mainly on the network size, *i.e.*, the total number of sensors in the WSN. This property is consistent for algorithms executed in a centralized manner [18], where the acquired information is conveyed to a central processor that performs the necessary computations. As shown in Table 2.1, in the case of non-cooperative localization, the proposed estimators based on GTRS framework are slightly more complex than the existing one, due to the iterative bisection procedure. However, the higher computational complexity of the proposed estimators is justified by their superior performance in the sense of estimation accuracy, as we will see in Section 2.2.6. Finally, from Table 2.1 we can see that the proposed

[1]In practice, however, the number of connections in the network is significantly smaller, due to energy restrictions, *i.e.*, limited R.

Table 2.1 Summary of the considered algorithms

Algorithm	Description	Complexity
SR-WLS1	The proposed SR-WLS estimator for non-cooperative localization when P_T is known (2.40)	$K_{\max} \cdot \mathcal{O}(N)$
LS	The LS estimator for non-cooperative localization when P_T is known in [30]	$\mathcal{O}(N)$
SR-WLS2	The proposed SR-WLS approach for non-cooperative localization in Section 2.2.3 for unknown P_T	$2 \cdot K_{\max} \cdot \mathcal{O}(N)$
SDP1	The proposed SDP estimator for cooperative localization when P_T is known in (2.21)	$\mathcal{O}\left(\sqrt{3M}\left(81M^4\left(N+\frac{M}{2}\right)^2\right)\right)$
SDP2	The proposed SDP approach for cooperative localization in Section 2.2.4 for unknown P_T	$2 \cdot \mathcal{O}\left(\sqrt{3M}\left(81M^4\left(N+\frac{M}{2}\right)^2\right)\right)$

estimators for cooperative localization are computationally the most demanding. This is not surprising, since the cooperative localization problem is very challenging, and requires the use of sophisticated mathematical tools in order to be solved globally.

2.2.6 Performance Results

In this section, we present a set of performance results to compare the proposed approaches with the existing ones, for both non-cooperative and cooperative localization with known and unknown P_T. In order to demonstrate the benefit of fusing two radio measurements versus traditional localization systems, we present also the performance results of the proposed methods for known P_T when only RSS measurements are employed, called here SR-WLS$_{\mathrm{RSS}}$ and SDP$_{\mathrm{RSS}}$ for non-cooperative and cooperative localization, respectively. Also, it is worth mentioning that for the sake of fairness, the range measurements for the LS method in [30], were acquired according to (2.2a). All of the presented algorithms were solved by using the MATLAB package `CVX` [71], where the solver is SeDuMi [73].

To generate the radio measurements, (2.1), (2.3) and (2.4) were used. We considered a random deployment of nodes inside a box with a length of the edges $B = 15$ m in each Monte Carlo (M_c) run. Random deployment of nodes is of practical interest, since the algorithms are tested against various network topologies. Unless stated otherwise, the reference distance is set to $d_0 = 1$ m, the reference power to $P_0 = -10$ dBm, the maximum number of steps in the bisection procedure to $K_{\max} = 30$, and the Path loss exponent (PLE) was fixed to $\gamma = 2.5$. However, in practice it is almost impossible to perfectly estimate the value of the PLE. Therefore, to account for a realistic measurement model mismatch and test the robustness of the considered approaches to imperfect knowledge of the PLE, the true PLE for each link was drawn from a uniform distribution on an interval $[2.2, 2.8]$, *i.e.*, $\gamma_{ij} \in [2.2, 2.8], \forall (i,j) \in \mathcal{A} \cup \mathcal{B}, i \neq j$.

2.2.6.1 Non-cooperative WSN

In a non-cooperative WSN, targets communicate exclusively with anchors and one target is located at a time; hence, without loss of generality, we can assume that $M = 1$. It was assumed that the RSS and AoA measurements were performed by anchors. As the main performance metric for the non-cooperative localization we used the root mean square error (RMSE), defined as

$$\text{RMSE} = \sqrt{\sum_{i=1}^{M_c} \frac{\|\boldsymbol{x}_i - \widehat{\boldsymbol{x}}_i\|^2}{M_c}},$$

where $\widehat{\boldsymbol{x}}_i$ denotes the estimate of the true target location, $\boldsymbol{x}_i$, in the i-th M_c run.

Figure 2.4 illustrates the RMSE versus N comparison when $\sigma_{n_{ij}} = 6$ dB, $\sigma_{m_{ij}} = 10$ deg and $\sigma_{v_{ij}} = 10$ deg. Besides the considered algorithms, we also present the CRB in the figure (derivation of the CRB for the RSS-AoA-based localization is presented in Appendix A) As anticipated, Figure 2.4 reveals that the performance of all algorithms improves as more anchors are added into the network, *i.e.*, as more reliable information is available. It also confirms the effectiveness of using the combined measurements in hybrid systems versus using only a single measurement like in traditional systems. Furthermore, one can see that the proposed estimators for both known and unknown P_T

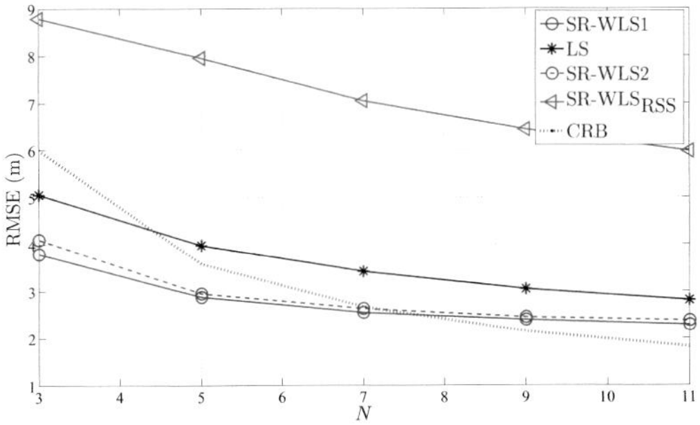

Figure 2.4 RMSE versus N comparison, when $\sigma_{n_{ij}} = 6$ dB, $\sigma_{m_{ij}} = 10$ deg, $\sigma_{v_{ij}} = 10$ deg, $\gamma_{ij} \in [2.2, 2.8]$, $\gamma = 2.5$, $B = 15$ m, $P_0 = -10$ dBm, $d_0 = 1$ m, $M_c = 50000$.

outperform significantly the existing one for all N. Additionally, it can be seen that "SR-WLS2" achieves the lower bound provided by its complement for known P_T, "SR-WLS1". We can also observe that the performance margin between the proposed estimators for known and unknown P_T decreases with the increase of N. This behaviour is intuitive, since with increased N we expect to obtain a better estimation of P_T (*closer* to its true value), which would allow us to enhance the estimation accuracy in the third step of our proposed procedure. Finally, although our estimators were based on the assumption that the noise is small, Figure 2.4 reveals that they work excellent even for the cases where the noise power is high.

In Figures 2.5, 2.6 and 2.7 we investigate the influence of the quality of certain types of measurements on the performance of the considered approaches. More specifically, Figures 2.5, 2.6 and 2.7 illustrate the RMSE versus $\sigma_{n_{ij}}$ (dB), $\sigma_{m_{ij}}$ (deg) and $\sigma_{v_{ij}}$ (deg) comparison when $N = 4$, respectively. From these figures, one can observe that when the quality of a certain measurement drops, the performance of the considered algorithms worsens, as expected. Further, one can see that both the proposed and the existing approach suffer the biggest deterioration in the performance when the quality of the RSS measurements weakens. Also, while the quality of the azimuth angle measurements affect more the proposed approaches than the "LS" method, the error in the elevation angle measurements has very little effect on their performance. Furthermore, it can be seen that the proposed procedure for unknown P_T is robust to noise feature, since the performance margin

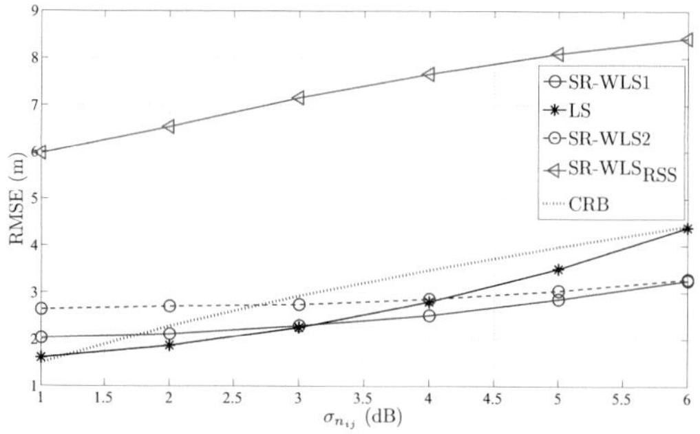

Figure 2.5 RMSE versus $\sigma_{n_{ij}}$ (dB) comparison, when $N = 4$, $\sigma_{m_{ij}} = 10$ deg, $\sigma_{v_{ij}} = 10$ deg, $\gamma_{ij} \in [2.2, 2.8]$, $\gamma = 2.5$, $B = 15$ m, $P_0 = -10$ dBm, $d_0 = 1$ m, $M_c = 50000$.

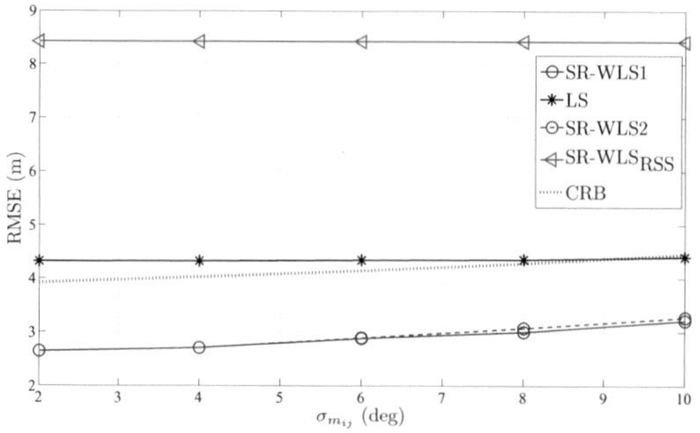

Figure 2.6 RMSE versus $\sigma_{m_{ij}}$ (deg) comparison, when $N = 4$, $\sigma_{n_{ij}} = 6$ dB, $\sigma_{v_{ij}} = 10$ deg, $\gamma_{ij} \in [2.2, 2.8]$, $\gamma = 2.5$, $B = 15$ m, $P_0 = -10$ dBm, $d_0 = 1$ m, $M_c = 50000$.

between "SR-WLS1" and its counterpart for unknown P_T remains constant with the increase of noise, in general. Finally, Figures 2.5, 2.6 and 2.7 exhibit superior performance of the proposed algorithms in comparison with the existing one, in general.

2.2.6.2 Cooperative WSN

This subsection presents the simulation results for the cooperative localization problem. As it was already mentioned, to the best of authors' knowledge, localization algorithms for hybrid RSS/AoA systems in cooperative 3-D WSNs are not available in the literature.

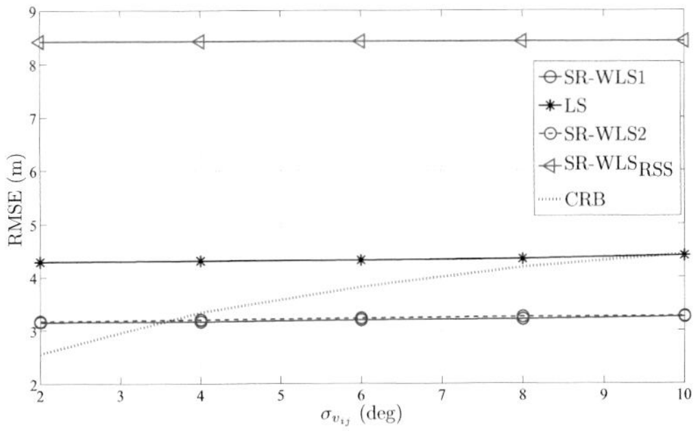

Figure 2.7 RMSE versus $\sigma_{v_{ij}}$ (deg) comparison, when $N = 4$, $\sigma_{n_{ij}} = 6$ dB, $\sigma_{m_{ij}} = 10$ deg, $\gamma_{ij} \in [2.2, 2.8]$, $\gamma = 2.5$, $B = 15$ m, $P_0 = -10$ dBm, $d_0 = 1$ m, $M_c = 50000$.

Therefore, only the performance of the proposed approaches in Sections 2.2.4 for both cases of known and unknown P_T are analysed. In this section, it was assumed that the signal measurements were performed by targets. As the main performance metric we used the normalized root mean square error (NRMSE), defined as

$$\text{NRMSE} = \sqrt{\sum_{i=1}^{M_c} \sum_{j=1}^{M} \frac{\|\boldsymbol{x}_{ij} - \widehat{\boldsymbol{x}}_{ij}\|^2}{M\, M_c}},$$

where $\widehat{\boldsymbol{x}}_{ij}$ denotes the estimate of the true location of the j-th target, $\boldsymbol{x}_{ij}$, in the i-th M_c run.

Figure 2.8 illustrates the NRMSE versus N comparison of the proposed estimators for both known and unknown P_T and the proposed SDP method for known P_T when only RSS measurements were used, for $M = 20$, $R = 8$ m, $\sigma_{n_{ij}} = 6$ dB, $\sigma_{m_{ij}} = 10$ deg and $\sigma_{v_{ij}} = 10$ deg. Figure 2.4 confirms that adding more reliable information into the network boosts the performance of all considered estimators, and decreases the performance margin between "SDP1" and "SDP2". This behavior is not unusual since by increasing N we are expected to obtain a better estimation of P_T (*closer* to its true value) in the second step of our proposed procedure which would enhance the estimation accuracy in the final step of the procedure. Furthermore, Figure 2.8 confirms that by fusing two radio measurements of the transmitted signal can

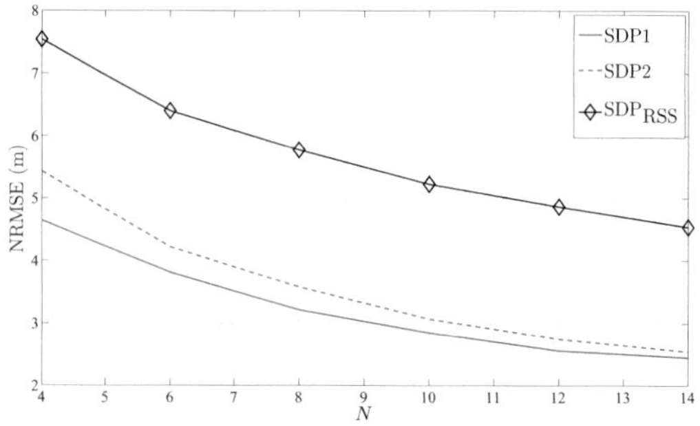

Figure 2.8 NRMSE versus N comparison, when $M = 20$, $R = 8$ m, $\sigma_{n_{ij}} = 6$ dB, $\sigma_{m_{ij}} = 10$ deg, $\sigma_{v_{ij}} = 10$ deg, $\gamma_{ij} \in [2.2, 2.8]$, $\gamma = 2.5$, $B = 15$ m, $P_0 = -10$ dBm, $d_0 = 1$ m, $M_c = 1000$.

significantly decrease the estimation error in comparison with using only one measurement.

Figure 2.9 illustrates the NRMSE versus M comparison of the considered estimators, when $N = 8$, $R = 8$ m, $\sigma_{n_{ij}} = 6$ dB, $\sigma_{m_{ij}} = 10$ deg and $\sigma_{v_{ij}} = 10$ deg. One can notice from Figure 2.9 that adding more targets into the network does not impair the performance of the considered estimators. In fact, their performance betters as M increases. It also exhibits that the performance margin between "SDP1" and "SDP2" slowly grows with increase of M. This might be explained by the fact that when more targets are added in the network, more unreliable measurements are obtained (set $\mathcal{B}$ is enlarged) which might

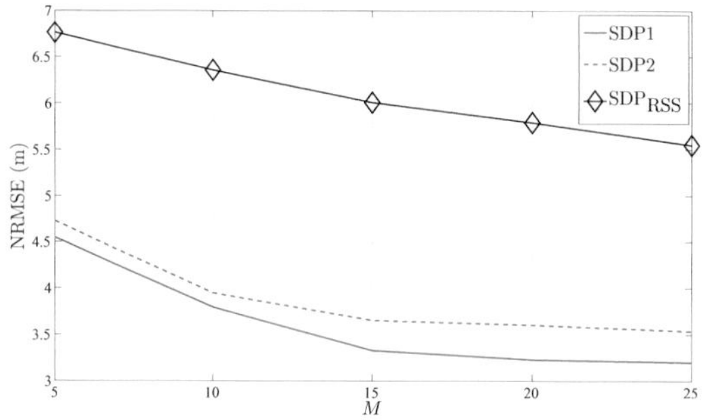

Figure 2.9 NRMSE versus M comparison, when $N = 8$, $R = 8$ m, $\sigma_{n_{ij}} = 6$ dB, $\sigma_{m_{ij}} = 10$ deg, $\sigma_{v_{ij}} = 10$ deg, $\gamma_{ij} \in [2.2, 2.8]$, $\gamma = 2.5$, $B = 15$ m, $P_0 = -10$ dBm, $d_0 = 1$ m, $M_c = 1000$.

deteriorate the estimation of P_T and consequently the location estimation. Finally, although the measurement noise is high in Figure 2.9, we can see that the proposed methods perform excellent.

Figure 2.10 illustrates the NRMSE versus R comparison of the considered estimators, when $N = 8$, $M = 20$, $\sigma_{n_{ij}} = 6$ dB, $\sigma_{m_{ij}} = 10$ deg and $\sigma_{v_{ij}} = 10$ deg. Figure 2.10 shows that the estimation error of the new estimators decreases as R increases. This behavior is anticipated, since when R grows, the acquired information inside the network also grows, as well as the probability that more target/anchor connections are established. One can observe that the performance margin between the hybrid methods and the RSS one increases as R grows. This is because when R is too low (*e.g.* $R = 5$ m) the amount of information obtained from sensors is insufficient, resulting in a poor estimation accuracy (NRMSE ≈ 8 m). Obviously, when R is expanded the proposed hybrid methods benefit more from the additional links than "SDP$_{\text{RSS}}$" method, since for each additional link two measurements (RSS and AoA) are performed. Note however, that increasing R directly impacts the sensor's battery life, and that in practice we want to keep R as low as possible[2].

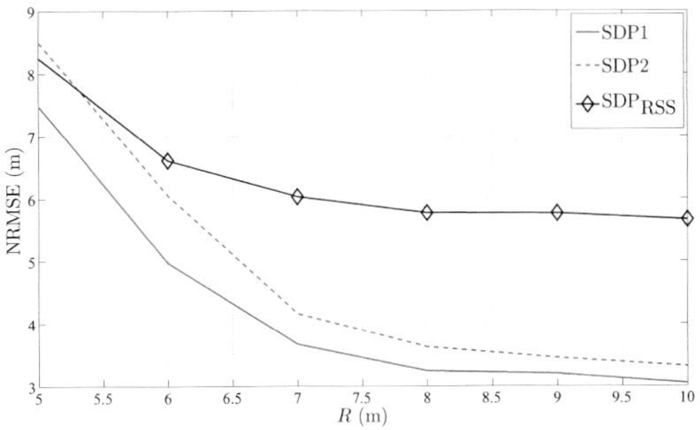

Figure 2.10 NRMSE versus R comparison, when $N = 8$, $M = 20$, $\sigma_{n_{ij}} = 6$ dB, $\sigma_{m_{ij}} = 10$ deg, $\sigma_{v_{ij}} = 10$ deg, $\gamma_{ij} \in [2.2, 2.8]$, $\gamma = 2.5$, $B = 15$ m, $P_0 = -10$ dBm, $d_0 = 1$ m, $M_c = 1000$.

[2]The estimators in [31, 32, 60] were not considered here, since they were designed for 2-D scenarios, and a possible generalization to a 3-D scenario is not obvious. However, it is worth mentioning that, in our simulations, the proposed algorithms

2.2.6.3 Real indoor experiment

In this section, we asses the performance of our SR-WLS2 algorithm that takes advantage of both RSS and AoA measurements through a real indoor experiment. Our experiment is based entirely on the measurements performed in [67]. Figure 2.11 illustrates the 56 m × 25 m building in which the measurements were taken. In the figure, the true locations of the measurement points (targets) are indicated by blue circles and the true locations of the base stations (anchors) are indicated by black squares.

In [67], a regular 802.11 equipped laptop took four sets of measurements at each measurement point, one for each pose of the target (facing north, east, south, and west). No compass was used for orientation, and the target was set just by aiming to have the measurement laptop parallel with the walls. The location of the target was randomized in order to include in the measurements situations where both a human body and a laptop screen block was the shortest path towards a base station. Although the polarization did not seem to matter, the authors in [67] performed all the measurements with 802.11 card kept in an horizontal plane, as is standard in most laptops. A measurement

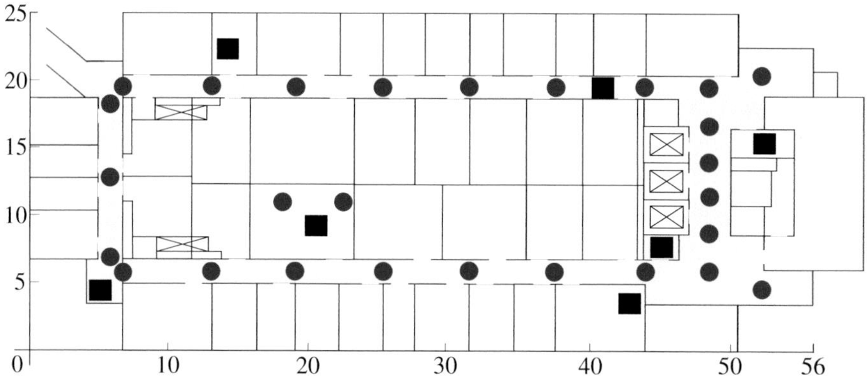

Figure 2.11 Experimental set-up with 7 anchors (black squares) and 27 targets (blue circles).

---

outperformed the mentioned ones in terms of the estimation accuracy in 2-D scenarios. Note also that the WLS estimator in [33] for 3-D scenarios was omitted here. The reason is that this estimator did not exhibit acceptable performance in the investigated settings, where the area accommodating nodes and the noise power are much larger than it was considered in [33].

for a pose was in fact an average over three or four revolutions of the base station, in order to reduce the effect of temporary factors, such as open doors, or people passing by. Measurements were taken at various times of the day and night, including the busy morning and afternoon hours.

The way the authors in [67] managed to extract the AoA measurements on a 802.11 base station was to attach a directional antenna to a wireless access point. When this antenna was rotated, the RSS reported by the card was higher in the direction of the measurement point in general, see Figure 2.12. To automatize this measurement of the angle, the authors mounted a small Toshiba Libretto 70ct laptop on a record player (turntable). In order to obtain higher difference in the maximums, they chose an antenna that is highly directional. The Lucent 2 Mbps 802.11 card was linked to a Hyperlink 14 dB gain directional antenna. The antenna was attached to the bottom of the laptop, so that it rotates in the horizontal plane.

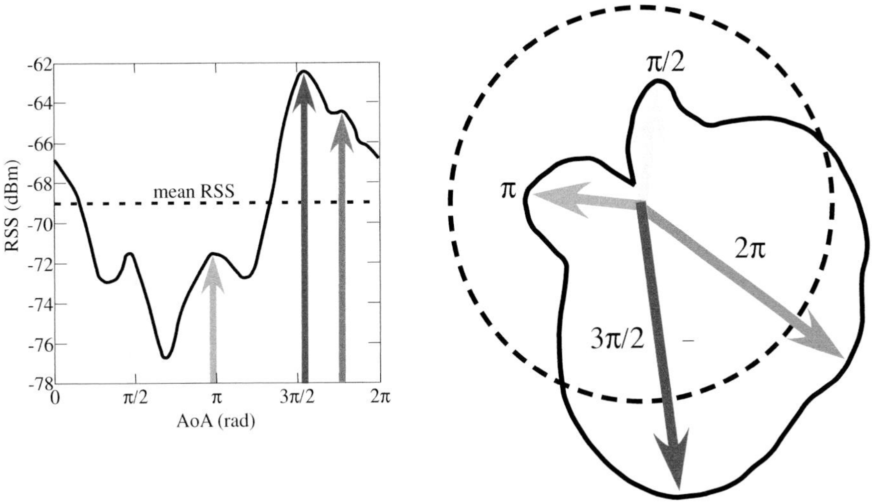

(a) Cartesian representation of RSS peak (b) Polar representation of RSS peak

Figure 2.12 Illustration of the RSS peaks, indicating a possible direction of the target.

In Figure 2.12, one can see that there is more than one peak in the RSS measurements, each one indicating a possible direction of the target. Obviously, choosing the right peak is a critical element of the measurement procedure, since failing to do so might result in large errors. However, the results in [67] indicate that the best measured AoA comes either from the first or the second RSS peak in 90% of the cases. Such results are not surprising in indoor environments, where a mixture of line-of-sight (LoS) and non-line-of-sight (NLoS) links exist, since in such environments it is expected to get one peak in the RSS from a signal propagating through a corridor and the other one from a signal passing through walls, the so-called quasi LoS [74]. Nevertheless, there still remain 10% of the cases in which the right direction comes from other RSS peaks. To resolve this issue, in this work, we have exploited our knowledge about the building configuration and the known locations of the anchors in order to eliminate those AoA measurements that do not make any sense. For example, it only made sense that the anchor physically closest to the origin and the one furthest away in the y-axis direction could only obtain AoA measurements in the I and IV, and in the III and IV quadrants of a Cartesian coordinate system, respectively. Therefore, if the first peak resulted in a direction outside these quadrants, we have disregarded it and chose the second one, and so on.

By using the measurements in [67], we first applied a linear regression method to obtain $\gamma = 3.4$. For such an experimental setup, we compared the performance of our SR-WLS2 algorithm with the LS method used in [67]. We present the cumulative distribution of the localization error (LE), where LE $= \|\boldsymbol{x}_i - \widehat{\boldsymbol{x}}_i\|$, for $i = 1, ..., M_c$, in Figure 2.13 for different number of *closest* anchors utilized. The figure shows that as N is increased, the performance of both methods betters in general, as anticipated. Still, the figure shows that the obtained results are not always better for larger N. Such a behaviour can be explained to some extent by the fact that the experimental setup perhaps was not ideally balanced in the sense of anchors' locations, resulting in considerable difference between links' length. This might produce *bad* links (very distant from the source), which, when taken into consideration, can have negative impact on the estimation accuracy of an algorithm. Furthermore, one can see that both estimators perform well in the considered scenario, since the median error for $N = 7$ is pretty much the same for both estimators, just above 2 m. However, it can be seen that the performance of the SR-WLS2 is more stable, as it produces an LE ≤ 4 m in more than 80% of the cases. This is somewhat expected and can be explained to some extent by the fact that our

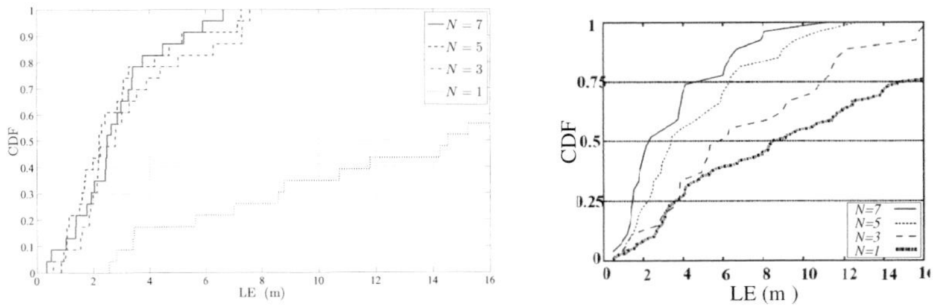

(a) SR-WLS2 (b) The approach in [67]

Figure 2.13 Cumulative distribution of the localization error.

SR-WLS2 method is executed iteratively, which raises its computational complexity. Nevertheless, Figure 2.13 exhibits a superior performance of our SR-WLS2 method for low N, in general.

In favor of testing the hypothesis that hybrid methods perform better than *traditional* ones, we present also the RMSE (m) versus N performance comparison of our SR-WLS2 estimator in the case where we used RSS only and combined RSS and AoA measurements in Figure 2.14. It can be seen that the performance of both approaches betters when N grows. Furthermore, the figure confirms the superiority of the hybrid approach, showing significant error reduction in the case where measurements are combined. However, the performance margin between the considered approaches reduces as N is increased, as expected. Nevertheless, the advantage of the hybrid approach over the *traditional* one is significant (more than 3 m) even for $N = 7$.

Finally, although the indoor localization scenario is very challenging, based on the results obtained in the considered indoor experiment, we can conclude that our method performs well in such a surrounding.

2.2.7 Conclusions

In this section, we addressed the hybrid RSS/AoA target localization problem in both non-cooperative and cooperative 3-D WSN, for both cases of known and unknown P_T. We first developed a novel non-convex objective function from the RSS and AoA measurement models. For the case of non-cooperative localization, we showed that the derived objective function can be transformed into a GTRS framework, by following the SR approach. Moreover, we showed that the derived non-convex

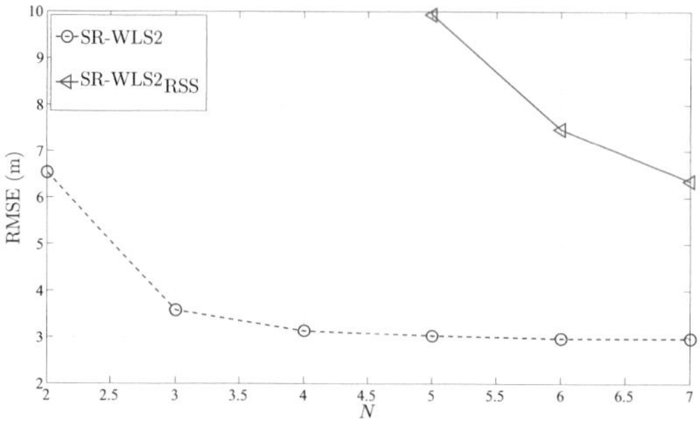

Figure 2.14 RMSE versus N performance comparison in the considered experimental setup.

objective function can be transformed into a convex one, by applying SDP relaxation technique in the case of cooperative localization. For the case where P_T is not known, we proposed a three-step procedure in order to enhance the estimation accuracy of our algorithms. The simulation results confirmed the effectiveness of the new algorithms in a variety of settings. For the case of non-cooperative localization, the simulation results show that the proposed approaches significantly outperform the existing one, even for the case where the proposed estimators have no knowledge about P_T. For the case of cooperative localization, we have investigated the influence of N, M and R on the estimation accuracy. For all considered scenarios, the new estimators exhibited excellent performance, and robustness to not knowing P_T.

2.3 Distributed RSS-AoA-based Target Localization

2.3.1 Related Work

Localization of a sensor network with small number of anchors using graph theory and binary data has drawn much attention recently [75, 76]. In [77] a study of traditional non-cooperative RSS- and AoA-based localization methods for visible light communication systems was presented. The approaches in [26–29] are based on the fusion of RSS and ToA measurements. A hybrid system that merges range and angle measurements was investigated in [30]. The authors in [30] proposed two estimators to solve the non-cooperative target localization problem in a 3-D scenario: linear LS and optimization based. The LS estimator is a relatively simple and well known estimator, while the optimization based estimator was solved by Davidon-Fletcher-Powell algorithm [59]. In [53], the authors derived an LS and a ML estimator for a hybrid scheme that combines RSSD and AoA measurements. Non-linear constrained optimization was used to estimate the target's location from multiple RSS and AoA measurements. Both LS and ML estimators in [53] are λ-dependent, where λ is a non-negative weight assigned to regulate the contribution from RSS and AoA measurements. A selective WLS estimator for RSS/AoA localization problem was proposed in [32]. The authors determined the target location by exploiting weighted ranges from the two *nearest* anchor measurements, which were combined with the serving base station AoA measurement. In [53] and [32], authors investigated the non-cooperative hybrid RSS/AoA localization problem for a 2-D scenario only. A WLS estimator for a 3-D RSSD/AoA non-cooperative localization problem when the transmit power is unknown was presented in [33]. However, the authors in [33] only investigated a small-scale WSN, with extremely low noise power. Two estimators for 3-D non-cooperative RSS/AoA localization problem based on convex optimization and squared-range approach were proposed in [34]. The work in [35] addressed an RSS/AoA non-cooperative localization problem in 2-D non-line of sight environments. The authors in [35] proposed an alternating optimization algorithm, composed of fixing the value of the scatter orientation and solving the SDP representation of the localization problem and later using the obtained location estimate to update the value of the scatter orientation, for localizing a mobile target in a WSN. In [36] a cooperative RSS/AoA localization problem was investigated. The authors in [36] proposed an SDP estimator to simultaneously localize multiple targets. However, the proposed algorithm is for centralized applications only, and its computational complexity depends highly on the network size. Convex optimization techniques were employed in [38] to solve the cooperative RSS/AoA target localization problem with unknown transmit powers in a distributed manner.

2.3.1.1 Contribution

Apart from [36] and [38], all mentioned approaches investigate non-cooperative localization problem only, where the location of a single target, which communicates with anchors exclusively, is determined at a time. Contrary to these approaches, in this work we investigate the target localization problem in a large-scale WSN, where the number of anchors is scarce and the communication range of all sensors is restricted (*e.g.*, to prolong sensor's battery life). In such settings, only some targets can directly communicate with anchors; therefore, cooperation between any two sensors within the communication range is required in order to acquire sufficient amount of information to perform localization. We design novel distributed hybrid localization algorithms based on SOCP relaxation and GTRS framework that take advantage of combined RSS/AoA measurements with known transmit power to estimate the locations of all targets in a WSN. The proposed algorithms are distributed in the sense that no central sensor coordinates the network, all communications occur exclusively between two incident sensors and the data associated with each sensor are processed locally. First, the non-convex and computationally complex ML estimation problem is broken down into smaller sub-problems, *i.e.*, the local ML estimation problem for each target is posed. By using the RSS propagation model and simple geometry, we derive a novel local non-convex estimator based on the LS criterion, which tightly approximates the local ML one for small noise levels. Then, we show that the derived non-convex estimator can be transformed into a convex SOCP estimator that can be solved efficiently by interior-point algorithms [56]. Furthermore, following the SR approach, we propose a suboptimal SR-WLS estimator based on the GTRS framework, which can be solved exactly by a bisection procedure [69]. We then generalize the proposed SOCP estimator for known transmit powers to the case where the target transmit powers are different and not known.

2.3.2 Problem Formulation

Consider a large-scale WSN with M targets and N anchors, randomly deployed over a region of interest. The considered network can be seen as a connected graph, $\mathcal{G}(\mathcal{V}, \mathcal{E})$, with $|\mathcal{V}| = M+N$ vertices and $|\mathcal{E}|$ edges, where $|\bullet|$ represents the cardinality (the number of elements in a set) of a set. The set of targets and the set of anchors are respectively labeled as $\mathcal{T}$ ($|\mathcal{T}| = M$) and $\mathcal{A}$ ($|\mathcal{A}| = N$), and their locations are denoted by $\boldsymbol{x}_1, \boldsymbol{x}_2, ..., \boldsymbol{x}_M$ and $\boldsymbol{a}_1, \boldsymbol{a}_2, ..., \boldsymbol{a}_N$ ($\boldsymbol{x}_i, \boldsymbol{a}_j \in \mathbb{R}^3$, $\forall i \in \mathcal{T}$ and $\forall j \in \mathcal{A}$), respectively. To save power (battery duration conditions the lifetime of a network), it is assumed that all sensors have limited communication range, R. Thus, two sensors, i and j, can exchange information if and only if they are within the communication range of each other. The sets of all target/anchor and target/target connections (edges) are defined as $\mathcal{E}_{\mathcal{A}} = \{(i,j) : \|\boldsymbol{x}_i - \boldsymbol{a}_j\| \leq R, \forall i \in \mathcal{T}, \forall j \in \mathcal{A}\}$ and $\mathcal{E}_{\mathcal{T}} = \{(i,k) : \|\boldsymbol{x}_i - \boldsymbol{x}_k\| \leq R, \forall i,k \in \mathcal{T}, i \neq k\}$, respectively.

For ease of expression, let us define a matrix $\boldsymbol{X} = [\boldsymbol{x}_1, \boldsymbol{x}_2, ..., \boldsymbol{x}_M]$ ($\boldsymbol{X} \in \mathbb{R}^{3\times M}$) as the matrix of all unknown target locations. We determine these locations by using a hybrid system that fuses range and angle measurements.

Throughout this work, it is assumed that the range measurements are obtained from the RSS information exclusively, since ranging based on RSS requires the lowest implementation costs [1]. The RSS between two sensors i and j which are within the communication range of each other (from the transmitting sensor), P_{ij} (dBm), is modeled as:

$$P_{ij}^{\mathcal{A}} = P_{0i} - 10\gamma \log_{10} \frac{\|\boldsymbol{x}_i - \boldsymbol{a}_j\|}{d_0} + n_{ij}, \forall (i,j) \in \mathcal{E}_{\mathcal{A}}, \tag{2.25a}$$

$$P_{ik}^{\mathcal{T}} = P_{0i} - 10\gamma \log_{10} \frac{\|\boldsymbol{x}_i - \boldsymbol{x}_k\|}{d_0} + n_{ik}, \forall (i,k) \in \mathcal{E}_{\mathcal{T}}, \tag{2.25b}$$

(see [61, 62]), where n_{ij} and n_{ik} are the log-normal shadowing terms modeled as $n_{ij} \sim \mathcal{N}(0, \sigma_{n_{ij}}^2)$, $n_{ik} \sim \mathcal{N}(0, \sigma_{n_{ik}}^2)$. We assume that the target/target RSS measurements are symmetric[1], *i.e.*, $P_{ik}^{\mathcal{T}} = P_{ki}^{\mathcal{T}}, \forall (i,k) \in \mathcal{E}_{\mathcal{T}}, i \neq k$.

To obtain the AoA measurements (both azimuth and elevation angles), we assume that either antenna arrays or a directional antenna is implemented at anchors [30, 66, 67, 78, 79], or that the anchors are equipped with video cameras [63–65]. In order to make use of the AoA measurements from different anchors, the orientation information is required, which can be obtained by implementing a digital compass at each anchor [30, 66]. However, a digital compass introduces an error in the AoA measurements due to its static accuracy. For the sake of simplicity and without loss of generality, we model the angle measurement error and the orientation error as one random variable in the rest of this work.

Figure 2.15 gives an illustration of a target and an anchor locations in a 3-D space. As shown in Figure 2.15, $\boldsymbol{x}_i = [x_{i\text{x}}, x_{iy}, x_{iz}]^T$ and

[1]This assumption is made without loss of generality; it is readily seen that, if $P_{ik}^{\mathcal{T}} \neq P_{ki}^{\mathcal{T}}$, then it is enough to replace $P_{ik}^{\mathcal{T}} \leftarrow (P_{ik}^{\mathcal{T}} + P_{ki}^{\mathcal{T}})/2$ and $P_{ki}^{\mathcal{T}} \leftarrow (P_{ik}^{\mathcal{T}} + P_{ki}^{\mathcal{T}})/2$ when solving the localization problem.

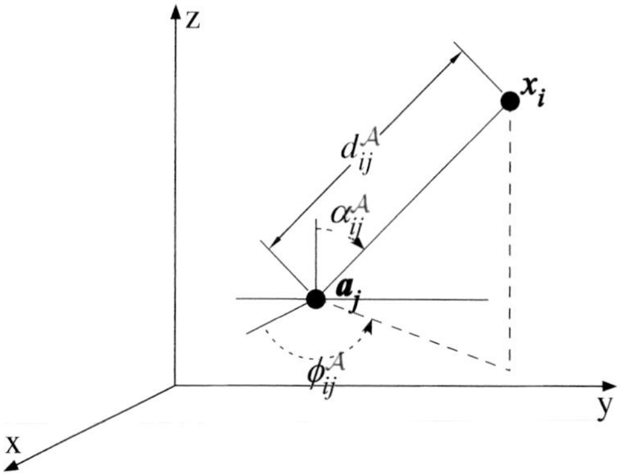

Figure 2.15 Illustration of a target and anchor locations in a 3-D space.

$\boldsymbol{a}_j = [a_{j\text{x}}, a_{j\text{y}}, a_{jz}]^T$ are respectively the unknown coordinates of the i-th target and the known coordinates of the j-th anchor, while $d_{ij}^{\mathcal{A}}$, $\phi_{ij}^{\mathcal{A}}$ and $\alpha_{ij}^{\mathcal{A}}$ represent the distance, azimuth angle and elevation angle between the i-th target and the j-th anchor, respectively. The ML estimate of the distance between two sensors can be obtained from the RSS measurement model (2.25) as follows [1]:

$$\widehat{d}_{ij} = \begin{cases} d_0 10^{\frac{P_{0i} - P_{ij}^{\mathcal{A}}}{10\gamma}}, & \text{if } j \in \mathcal{A}, \\ d_0 10^{\frac{P_{0i} - P_{ij}^{\mathcal{T}}}{10\gamma}}, & \text{if } j \in \mathcal{T}. \end{cases} \tag{2.26}$$

Applying simple geometry, azimuth and elevation angle measurements[2] can be modeled respectively as [30]:

$$\phi_{ij}^{\mathcal{A}} = \arctan\left(\frac{x_{i\text{y}} - a_{j\text{y}}}{x_{i\text{x}} - a_{j\text{x}}}\right) + m_{ij}, \text{ for } (i,j) \in \mathcal{E}_{\mathcal{A}}, \tag{2.27}$$

and

$$\alpha_{ij}^{\mathcal{A}} = \arccos\left(\frac{x_{i\text{z}} - a_{j\text{z}}}{\|\boldsymbol{x}_i - \boldsymbol{a}_j\|}\right) + v_{ij}, \text{ for } (i,j) \in \mathcal{E}_{\mathcal{A}}, \tag{2.28}$$

[2]Note that we consider here the case where only anchors have the necessary equipment to perform the respective angle measurements. An alternative approach would be to provide the necessary equipment to all sensors. However, our simulations showed that there is no gain for such a setting, and it would severely raise the overall network implementation costs.

where m_{ij} and v_{ij} are the measurement errors of azimuth and elevation angles, respectively, modeled as $m_{ij} \sim \mathcal{N}(0, \sigma^2_{m_{ij}})$ and $v_{ij} \sim \mathcal{N}(0, \sigma^2_{v_{ij}})$.

Given the observation vector $\boldsymbol{\theta} = [\boldsymbol{P}^T, \boldsymbol{\phi}^T, \boldsymbol{\alpha}^T]^T$ ($\boldsymbol{\theta} \in \mathbb{R}^{3|\mathcal{E}_\mathcal{A}|+|\mathcal{E}_\mathcal{T}|}$), where $\boldsymbol{P} = [P^\mathcal{A}_{ij}, P^\mathcal{T}_{ik}]^T$, $\boldsymbol{\phi} = [\phi^\mathcal{A}_{ij}]^T$, $\boldsymbol{\alpha} = [\alpha^\mathcal{A}_{ij}]^T$, the PDF is given as:

$$p(\boldsymbol{\theta}|\boldsymbol{X}) = \prod_{i=1}^{3|\mathcal{E}_\mathcal{A}|+|\mathcal{E}_\mathcal{T}|} \frac{1}{\sqrt{2\pi\sigma_i^2}} \exp\left\{ -\frac{(\theta_i - f_i(\boldsymbol{X}))^2}{2\sigma_i^2} \right\}, \tag{2.29}$$

where

$$\boldsymbol{f}(\boldsymbol{X}) = \begin{bmatrix} \vdots \\ P_{0i} - 10\gamma \log_{10} \frac{\|\boldsymbol{x}_i - \boldsymbol{a}_j\|}{d_0} \\ \vdots \\ P_{0i} - 10\gamma \log_{10} \frac{\|\boldsymbol{x}_i - \boldsymbol{x}_k\|}{d_0} \\ \vdots \\ \arctan\left(\frac{x_{iy} - a_{jy}}{x_{ix} - a_{jx}}\right) \\ \vdots \\ \arccos\left(\frac{x_{iz} - a_{jz}}{\|\boldsymbol{x}_i - \boldsymbol{a}_j\|}\right) \\ \vdots \end{bmatrix}, \quad \boldsymbol{\sigma} = \begin{bmatrix} \vdots \\ \sigma_{n_{ij}} \\ \vdots \\ \sigma_{n_{ik}} \\ \vdots \\ \sigma_{m_{ij}} \\ \vdots \\ \sigma_{v_{ij}} \\ \vdots \end{bmatrix}.$$

Maximizing the log of the likelihood function (2.29) with respect to $\boldsymbol{X}$ gives us the ML estimate, $\hat{\boldsymbol{X}}$, of the unknown locations [46], as:

$$\hat{\boldsymbol{X}} = \arg\min_{\boldsymbol{X}} \sum_{i=1}^{3|\mathcal{E}_\mathcal{A}|+|\mathcal{E}_\mathcal{T}|} \frac{1}{\sigma_i^2} \left[\theta_i - f_i(\boldsymbol{X})\right]^2. \tag{2.30}$$

Asymptotically (for large data records) the ML estimator in (2.30) is the minimum variance unbiased estimator [46]. However, finding the ML estimate directly from (2.30) is not possible, since (2.30) is non-convex and has no closed-form solution. Nevertheless, in the remainder of this work we will show that the LS problem in (2.30) can be solved in a distributed manner by applying certain approximations. More precisely, we propose a convex relaxation technique leading to a distributed SOCP estimator that can be solved efficiently by interior-point algorithms [56], and a suboptimal estimator based on the GTRS framework leading to a distributed SR-WLS estimator, which can be solved *exactly* by a bisection procedure [69]. We also show that the proposed SOCP estimator can be generalized to solve the localization problem in (2.30) where, besides the target locations, their transmit powers are different and unknown.

2.3.2.1 Assumptions

We outline here some assumptions for the WSN (made for the sake of simplicity and without loss of generality):

(1) The network is connected and it does not change during the computation period;
(2) Measurement errors for RSS and AoA models are independent, and $\sigma_{n_{ij}} = \sigma_n$, $\sigma_{m_{ij}} = \sigma_m$ and $\sigma_{v_{ij}} = \sigma_v$, $\forall (i,j) \in \mathcal{E}_\mathcal{A} \cup \mathcal{E}_\mathcal{T}$;
(3) The necessary equipment for collecting the AoA measurements is installed at anchors exclusively;
(4) A coloring scheme of the network is available.

In assumption (1), we assume that the sensors are static and that there is no sensor/link failure during the computation period, and that there exists a path between any two sensors $i, j \in \mathcal{V}$. Assumption (2) is made for the sake of simplicity. Assumption (3) indicates that only anchors are suitably equipped to acquire the AoA measurements (*e.g.* with directional antenna or antenna array [30, 66, 78], or video cameras [63, 64]), due to network costs. Finally, assumption (4) implies that a coloring scheme is available in order to color (number) the sensors and establish a working hierarchy in the network. More precisely, we assume that a second-order coloring scheme is employed, meaning that no sensor has the same color (number) as any of its one-hop neighbors nor its two-hop neighbors [23, 80, 81]. In this way, we avoid message collision and reduce the execution time of the algorithm, since sensors with the same color can work in parallel.

2.3.3 Distributed Localization

Notice that the problem in (2.30) is dependent on the locations and pairwise measurements between the adjacent sensors only. Thus, having the initial location estimations of the targets, $\hat{\boldsymbol{X}}^{(0)}$, at hand, the problem in (2.30) can be divided, *i.e.*, the minimization can be performed independently by each target using only the information gathered from its neighbors. Hence, rather than solving (2.30), which can be computationally exhausting (in large-scale WSNs), we break down (2.30) into sub-problems, which we solve locally (by each target) using iterative approach. Consequently, target i updates its location estimate in each iteration, t, by solving the following local ML problem:

$$\hat{\boldsymbol{x}}_i^{(t+1)} = \arg\min_{\boldsymbol{x}_i} \sum_{j=1}^{3|\mathcal{E}_{\mathcal{A}_i}|+|\mathcal{E}_{\mathcal{T}_i}|} \frac{1}{\sigma_j^2} \left[\theta_j - f_j(\boldsymbol{x}_i)\right]^2, \ \forall i \in \mathcal{T}, \tag{2.31}$$

where $\mathcal{E}_{\mathcal{A}_i} = \{j : (i,j) \in \mathcal{E}_{\mathcal{A}}\}$ and $\mathcal{E}_{\mathcal{T}_i} = \{k : (i,k) \in \mathcal{E}_{\mathcal{T}}, i \neq k\}$ represent the set of all anchor and all target neighbors of the target i respectively, and the first $|\mathcal{E}_{\mathcal{A}_i}| + |\mathcal{E}_{\mathcal{T}_i}|$ elements of $f_j(\boldsymbol{x}_i)$ are given as:

$$f_j(\boldsymbol{x}_i) = P_{0i} - 10\gamma \log_{10} \frac{\|\boldsymbol{x}_i - \hat{\boldsymbol{a}}_j\|}{d_0}, \text{ for } j = 1, ..., |\mathcal{E}_{\mathcal{A}_i}| + |\mathcal{E}_{\mathcal{T}_i}|,$$

with

$$\hat{\boldsymbol{a}}_j = \begin{cases} \boldsymbol{a}_j, & \text{if } j \in \mathcal{A}, \\ \hat{\boldsymbol{x}}_j^{(t)}, & \text{if } j \in \mathcal{T}. \end{cases}$$

2.3.3.1 Transmit powers are known

Distributed SOCP Algorithm. Assuming that $\hat{\boldsymbol{X}}^{(0)}$ is given, when the noise power is sufficiently small, from (2.25) we can write:

$$\lambda_{ij}\|\boldsymbol{x}_i - \hat{\boldsymbol{a}}_j\| \approx d_0, \ \forall i \in \mathcal{T}, \forall j \in \mathcal{E}_{\mathcal{A}_i} \cup \mathcal{E}_{\mathcal{T}_i}, \tag{2.32}$$

where

$$\lambda_{ij} = \begin{cases} 10^{\frac{P_{ij}^{\mathcal{A}} - P_{0i}}{10\gamma}}, & \text{if } j \in \mathcal{A}, \\ 10^{\frac{P_{ij}^{\mathcal{T}} - P_{0i}}{10\gamma}}, & \text{if } j \in \mathcal{T}. \end{cases}$$

Similarly, from (2.27) and (2.28) we respectively get:

$$\boldsymbol{c}_{ij}^T(\boldsymbol{x}_i - \boldsymbol{a}_j) \approx 0, \ \forall i \in \mathcal{T}, \forall j \in \mathcal{E}_{\mathcal{A}_i} \tag{2.33}$$

and

$$\boldsymbol{k}_{ij}^T(\boldsymbol{x}_i - \boldsymbol{a}_j) \approx \|\boldsymbol{x}_i - \boldsymbol{a}_j\| \cos(\alpha_{ij}^{\mathcal{A}}), \ \forall i \in \mathcal{T}, \forall j \in \mathcal{E}_{\mathcal{A}_i} \tag{2.34}$$

where $\boldsymbol{c}_{ij} = [-\sin(\phi_{ij}^{\mathcal{A}}), \cos(\phi_{ij}^{\mathcal{A}}), 0]^T$ and $\boldsymbol{k}_{ij} = [0, 0, 1]^T$. According to the LS criterion and (2.32), (2.33) and (2.34) each target updates its location by solving the following problem:

$$\hat{\boldsymbol{x}}_i^{(t+1)} = \arg\min_{\boldsymbol{x}_i} \sum_{j \in \mathcal{E}_{\mathcal{A}_i} \cup \mathcal{E}_{\mathcal{T}_i}} (\lambda_{ij} \|\boldsymbol{x}_i - \hat{\boldsymbol{a}}_j\| - d_0)^2$$
$$+ \sum_{j \in \mathcal{E}_{\mathcal{A}_i}} \left(\boldsymbol{c}_{ij}^T (\boldsymbol{x}_i - \boldsymbol{a}_j) \right)^2 + \sum_{j \in \mathcal{E}_{\mathcal{A}_i}} \left(\boldsymbol{k}_{ij}^T (\boldsymbol{x}_i - \boldsymbol{a}_j) - \|\boldsymbol{x}_i - \boldsymbol{a}_j\| \cos(\alpha_{ij}^{\mathcal{A}}) \right)^2 . \tag{2.35}$$

The LS problem in (2.35) is non-convex and has no closed-form solution. To convert (2.35) into a convex problem, we introduce auxiliary variables $r_{ij} = \|\boldsymbol{x}_i - \hat{\boldsymbol{a}}_j\|, \forall (i,j) \in \mathcal{E}_{\mathcal{A}} \cup \mathcal{E}_{\mathcal{T}}$, $\boldsymbol{z} = [z_{ij}]$, $\boldsymbol{g} = [g_{ij}]$, $\boldsymbol{p} = [p_{ij}]$, where $z_{ij} = \lambda_{ij}^{\mathcal{A}} r_{ij} - d_0, \forall (i,j) \in \mathcal{E}_{\mathcal{A}} \cup \mathcal{E}_{\mathcal{T}}$, $g_{ij} = \boldsymbol{c}_{ij}^T (\boldsymbol{x}_i - \boldsymbol{a}_j)$, and $p_{ij} = \boldsymbol{k}_{ij}^T (\boldsymbol{x}_i - \boldsymbol{a}_j) - r_{ij} \cos(\alpha_{ij}^{\mathcal{A}}), \forall (i,j) \in \mathcal{E}_{\mathcal{A}}$. We get:

$$\underset{\boldsymbol{x}_i, \boldsymbol{r}, \boldsymbol{z}, \boldsymbol{g}, \boldsymbol{p}}{\text{minimize}} \quad \|\boldsymbol{z}\|^2 + \|\boldsymbol{g}\|^2 + \|\boldsymbol{p}\|^2$$

subject to

$$\begin{aligned} r_{ij} &= \|\boldsymbol{x}_i - \hat{\boldsymbol{a}}_j\|, \ \forall (i,j) \in \mathcal{E}_{\mathcal{A}} \cup \mathcal{E}_{\mathcal{T}}, \\ z_{ij} &= \lambda_{ij} r_{ij} - d_0, \ \forall (i,j) \in \mathcal{E}_{\mathcal{A}} \cup \mathcal{E}_{\mathcal{T}}, \\ g_{ij} &= \boldsymbol{c}_{ij}^T (\boldsymbol{x}_i - \boldsymbol{a}_j), \ \forall (i,j) \in \mathcal{E}_{\mathcal{A}}, \\ p_{ij} &= \boldsymbol{k}_{ij}^T (\boldsymbol{x}_i - \boldsymbol{a}_j) - r_{ij} \cos(\alpha_{ij}^{\mathcal{A}}), \ \forall (i,j) \in \mathcal{E}_{\mathcal{A}}. \end{aligned} \tag{2.36}$$

Introduce epigraph variables e_1, e_2 and e_3, and apply second-order cone constraint relaxation of the form $\|\boldsymbol{z}\|^2 \leq e_1$, to obtain:

$$\underset{\boldsymbol{x}_i, \boldsymbol{r}, \boldsymbol{z}, \boldsymbol{g}, \boldsymbol{p}, e_1, e_2, e_3}{\text{minimize}} \quad e_1 + e_2 + e_3$$

subject to

$$\begin{aligned} \|\boldsymbol{x}_i - \hat{\boldsymbol{a}}_j\| &\leq r_{ij}, \ \forall (i,j) \in \mathcal{E}_{\mathcal{A}} \cup \mathcal{E}_{\mathcal{T}}, \\ z_{ij} &= \lambda_{ij} r_{ij} - d_0, \ \forall (i,j) \in \mathcal{E}_{\mathcal{A}} \cup \mathcal{E}_{\mathcal{T}}, \\ g_{ij} &= \boldsymbol{c}_{ij}^T (\boldsymbol{x}_i - \boldsymbol{a}_j), \ \forall (i,j) \in \mathcal{E}_{\mathcal{A}}, \\ p_{ij} &= \boldsymbol{k}_{ij}^T (\boldsymbol{x}_i - \boldsymbol{a}_j) - r_{ij} \cos(\alpha_{ij}^{\mathcal{A}}), \ \forall (i,j) \in \mathcal{E}_{\mathcal{A}}, \\ \left\| \begin{bmatrix} 2\boldsymbol{z} \\ e_1 - 1 \end{bmatrix} \right\| &\leq e_1 + 1, \left\| \begin{bmatrix} 2\boldsymbol{g} \\ e_2 - 1 \end{bmatrix} \right\| \leq e_2 + 1, \left\| \begin{bmatrix} 2\boldsymbol{p} \\ e_3 - 1 \end{bmatrix} \right\| \leq e_3 + 1. \end{aligned} \tag{2.37}$$

The problem in (2.37) is an SOCP problem, which can be efficiently solved by the `CVX` package [71] for specifying and solving convex programs. In the further text, we will refer to (2.37) as "SOCP".

Distributed SR-WLS Algorithm. We can rewrite (2.32) as:

$$\lambda_{ij}^2 \|\boldsymbol{x}_i - \hat{\boldsymbol{a}}_j\|^2 \approx d_0^2, \ \forall (i,j) \in \mathcal{E}_\mathcal{A} \cup \mathcal{E}_\mathcal{T}. \tag{2.38}$$

In order to give more importance to the nearby links, introduce weights, $\boldsymbol{w} = [\sqrt{w_{ij}}]$, where

$$w_{ij} = 1 - \frac{\widehat{d}_{ij}}{\sum_{(i,j)\in\mathcal{E}_\mathcal{A}\cup\mathcal{E}_\mathcal{T}} \widehat{d}_{ij}}.$$

In (2.34), substitute $\|\boldsymbol{x}_i - \hat{\boldsymbol{a}}_j\|$ with $\widehat{d}_{ij}$ described in (2.26). According to the WLS criterion and (2.38), (2.33) and (2.34) each target updates its location by solving the following problem:

$$\hat{\boldsymbol{x}}_i^{(t+1)} = \arg\min_{\boldsymbol{x}_i} \sum_{j\in\mathcal{E}_{\mathcal{A}_i}\cup\mathcal{E}_{\mathcal{T}_i}} w_{ij}\left(\lambda_{ij}^2\|\boldsymbol{x}_i - \hat{\boldsymbol{a}}_j\|^2 - d_0^2\right)^2$$
$$+ \sum_{j\in\mathcal{E}_{\mathcal{A}_i}} w_{ij}\left(\boldsymbol{c}_{ij}^T(\boldsymbol{x}_i - \boldsymbol{a}_j)\right)^2 + \sum_{j\in\mathcal{E}_{\mathcal{A}_i}} w_{ij}\left(\boldsymbol{k}_{ij}^T(\boldsymbol{x}_i - \boldsymbol{a}_j) - \widehat{d}_{ij}\cos(\alpha_{ij}^\mathcal{A})\right)^2. \tag{2.39}$$

The above WLS estimator is non-convex and has no closed-form solution. However, we can express (2.39) as a quadratic programming problem whose *global* solution can be computed efficiently [69]. Using the substitution $\boldsymbol{y}_i = [\boldsymbol{x}_i^T, \|\boldsymbol{x}_i\|^2]^T, \forall i \in \mathcal{T}$, (2.39) can be rewritten as:

$$\hat{\boldsymbol{y}}_i^{(t+1)} = \arg\min_{\boldsymbol{y}_i} \|\boldsymbol{W}(\boldsymbol{A}\boldsymbol{y}_i - \boldsymbol{b})\|^2$$

subject to

$$\boldsymbol{y}_i^T \boldsymbol{D} \boldsymbol{y}_i + 2\boldsymbol{l}^T \boldsymbol{y}_i = 0, \tag{2.40}$$

where $\boldsymbol{W} = \operatorname{diag}\left(\left[w_{ij\in\mathcal{E}_{\mathcal{A}_i}\cup\mathcal{E}_{\mathcal{T}_i}}, w_{ij\in\mathcal{E}_{\mathcal{A}_i}}, w_{ij\in\mathcal{E}_{\mathcal{A}_i}}\right]\right)$,

$$\boldsymbol{A} = \begin{bmatrix} \vdots & \vdots \\ -2\lambda_{ij}^2 \hat{\boldsymbol{a}}_j^T & \lambda_{ij}^2 \\ \vdots & \vdots \\ \boldsymbol{c}_{ij}^T & 0 \\ \vdots & \vdots \\ \boldsymbol{k}_{ij}^T & 0 \\ \vdots & \vdots \end{bmatrix}, \ \boldsymbol{b} = \begin{bmatrix} \vdots \\ d_0^2 - \lambda_{ij}^2\|\hat{\boldsymbol{a}}_j\|^2 \\ \vdots \\ \boldsymbol{c}_{ij}^T \boldsymbol{a}_j \\ \vdots \\ \boldsymbol{k}_{ij}^T \boldsymbol{a}_j + \widehat{d}_{ij}^\mathcal{A}\cos(\alpha_{ij}^\mathcal{A}) \\ \vdots \end{bmatrix},$$

$$\boldsymbol{D} = \begin{bmatrix} \boldsymbol{I}_3 & \boldsymbol{0}_{3\times1} \\ \boldsymbol{0}_{1\times3} & 0 \end{bmatrix}, \boldsymbol{l} = \begin{bmatrix} \boldsymbol{0}_{3\times1} \\ -1/2 \end{bmatrix},$$

i.e., $\boldsymbol{W} \in \mathbb{R}^{3|\mathcal{E}_{\mathcal{A}_i}|+|\mathcal{E}_{\mathcal{T}_i}|\times 3|\mathcal{E}_{\mathcal{A}_i}|+|\mathcal{E}_{\mathcal{T}_i}|}$, $\boldsymbol{A} \in \mathbb{R}^{3|\mathcal{E}_{\mathcal{A}_i}|+|\mathcal{E}_{\mathcal{T}_i}|\times 4}$ and $\boldsymbol{b} \in \mathbb{R}^{3|\mathcal{E}_{\mathcal{A}_i}|+|\mathcal{E}_{\mathcal{T}_i}|\times 1}$.

The objective function and the constraint in (2.40) are both quadratic. This type of problem is known as GTRS [69, 70], and it can be solved *exactly* by a bisection procedure [69]. We denote (2.40) as "SR-WLS" in the remaining text.

In summary, the derivation of the above approaches can be described in two parts. In the first part, the local non-convex ML estimator in (2.31) is approximated by a different non-convex estimator, (2.35) and (2.39) respectively. The use of the objective functions in (2.35) and (2.39) is motivated by the fact that we get a much smoother surface in comparison to (2.31), at a cost of introducing some bias with respect to the ML solution (see Figure 2.16). If the bias effect is small, we might reach the ML solution by employing a local search around the solution of (2.35) and (2.39). In the second part of our approach, we convert (2.35) and (2.39) into a convex problem and GTRS framework, by following the above procedures.

Figure 2.16 illustrates a realization of the objective function in (2.31), for the case where the true sensors' locations were used and a realization of (2.35) and (2.39) after only one iteration, and (2.39) after three iterations, where the estimated targets' locations were used. The i-th target was located at $[2.0; 3.3]$, and it could directly communicate with its three anchor and three target neighbors. The noise standard deviation (STD) of RSS measurements was set to $\sigma_{n_{ij}} = 2$ dB and the noise STD of angle measurements was set to $\sigma_{m_{ij}} = 3$ deg, and the rest of the parameters follow the set-up described in Section 2.3.5. On the one hand, in Figure 2.16(a), where the true sensors' locations were used, one can see that the objective function is highly non-convex and its global minimum is located at $[2.4; 3.5]$. Due to non-convexity of the problem, recursive algorithms, such as gradient search method, might get trapped into a local minimum, causing large error in the location estimation process. On the other hand, in Figures 2.16(b), 2.16(c) and 2.16(d), where estimated targets' locations (obtained by solving the proposed "SOCP" and "SR-WLS" algorithm, respectively) were used, it can be seen that these objective functions are much smoother than the one in (2.31), and that the global minimum after only one iteration is

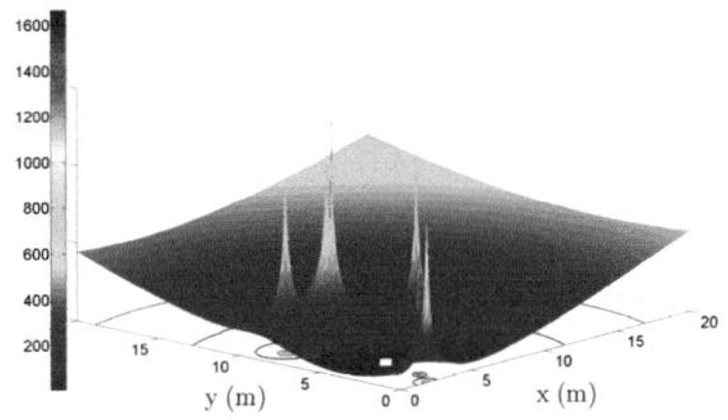

(a) Objective function in (2.31) using true sensors' locations

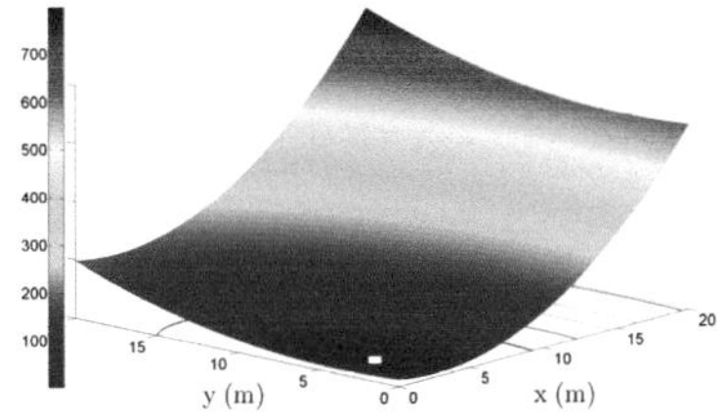

(b) Objective function in (2.35) after one iteration

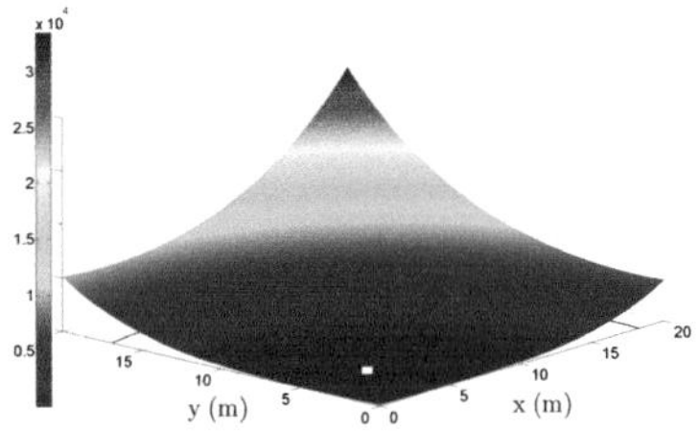

(c) Objective function in (2.39) after one iteration

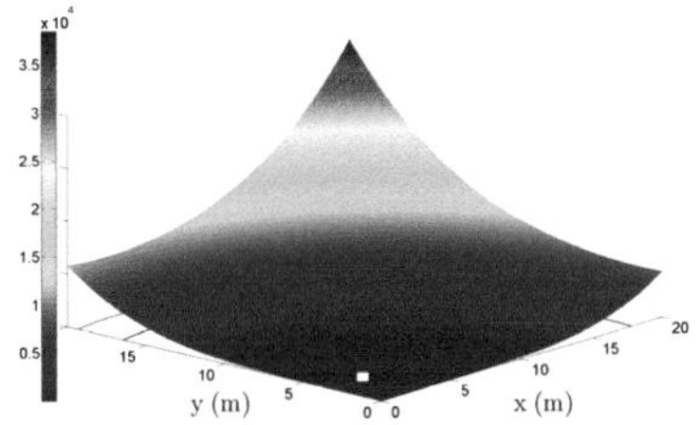

(d) Objective function in (2.39) after three iterations

Figure 2.16 Illustration of the objective functions in (2.31), (2.35) and (2.39) versus x (m) and y (m) coordinates (target location); the minimum of the objective function is indicated by a white square.

located at $[2.5; 4.1]$ and $[4.3; 4.9]$ for (2.35) and (2.39), respectively and at $[2.4; 3.6]$ for (2.39) after three iterations. Because of the smoothness of the objective functions, the global minimum of the considered problems can be obtained uniquely and effortlessly for all targets via interior-point algorithms [56] and bisection procedure [69], by following the proposed procedures. However, the *quality* of the obtained solution will depend on the tightness of the performed relaxation. As we show in Section 2.3.5, the estimation accuracy betters as the number of iterations grows in general. Thus, we can conclude that the objective functions in (2.35) and (2.39) represent an excellent approximation of the original problem defined in (2.31).

Assuming that $\mathcal{C}$ represents the set of colors of the sensors, Algorithm 1 summarizes the proposed distributed SOCP and SR-WLS algorithms. Algorithm 1 is distributed in the sense that there is no central processor in the network, its coordination is carried out according to the applied coloring scheme, information exchange occurs between two incident sensors exclusively, and data processing is performed locally by each target. Lines 5–7 are executed simultaneously by all targets $i \in \mathcal{C}_c$, which may decrease the execution time of the algorithm. At Line 6, we solve (2.37) if SOCP algorithm is employed, and (2.40) if SR-WLS algorithm is employed. The only information exchange occurs at Line 7, when targets broadcast their location updates $\hat{\boldsymbol{x}}_i^{(t+1)}$ to their neighbors. Since $\hat{\boldsymbol{x}}_i^{(t+1)} \in \mathbb{R}^3$, we can conclude that the proposed algorithm requires at most a broadcast of $3 \times T_{\max} \times M$ real values. Depending on which estimator is employed, in the remaining text, we label Algorithm 1 either as "SOCP" or as "SR-WLS".

Algorithm 1 The proposed distributed SOCP/SR-WLS algorithm

Require: $\hat{\boldsymbol{X}}^{(0)}$, $T_{\max}$, $\mathcal{C}$, $\boldsymbol{a}_j$, $\forall j \in \mathcal{A}$
1: **Initialize:** $t \leftarrow 0$
2: **repeat**
3: **for** $c = 1, ..., \mathcal{C}$ **do**
4: **for all** $i \in \mathcal{C}_c$ (in parallel) **do**
5: Collect $\hat{\boldsymbol{a}}_j, \forall j \in \mathcal{E}_{\mathcal{A}_i} \cup \mathcal{E}_{\mathcal{T}_i}$
6: $\hat{\boldsymbol{x}}_i^{(t+1)} \leftarrow \begin{cases} \text{solve (2.37),if using SOCP algorithm,} \\ \text{solve (2.40),if using SR-WLS algorithm} \end{cases}$
7: Broadcast $\hat{\boldsymbol{x}}_i^{(t+1)}$ to $\hat{\boldsymbol{a}}_j, \forall j \in \mathcal{E}_{\mathcal{A}_i} \cup \mathcal{E}_{\mathcal{T}_i}$
8: **end for**
9: **end for**
10: $t \leftarrow t + 1$
11: **until** $t < T_{\max}$

2.3.3.2 Transmit powers are not known

Often in practice testing and calibration are not the priority in order to restrict the implementation costs. Moreover, due to battery exhaust over time, sensors' transmit powers, P_i's, might change over time. Therefore, P_i's are often not calibrated, *i.e.*, not known. Not knowing P_i implies that P_{0i} is not known in the RSS model (2.25); see [16] and the references therein.

The generalization of the proposed SOCP estimator for known P_{0i} is straightforward for the case where P_{0i} is not known. More specifically, we can rewrite (2.32) as follows:

$$\zeta_{ij}\|\boldsymbol{x}_i - \hat{\boldsymbol{a}}_j\| \approx \eta_i d_0, \forall i \in \mathcal{T}, \forall j \in \mathcal{E}_{\mathcal{A}} \cup \mathcal{E}_{\mathcal{T}}, \tag{2.41}$$

where $\eta_i = 10^{\frac{P_{0i}}{10\gamma}}$ and

$$\zeta_{ij} = \begin{cases} 10^{\frac{P_{ij}^{\mathcal{A}}}{10\gamma}}, & \text{if } j \in \mathcal{A}, \\ 10^{\frac{P_{ij}^{\mathcal{T}}}{10\gamma}}, & \text{if } j \in \mathcal{T}. \end{cases}$$

Following the LS concept and (2.41), (2.33) and (2.34), each target updates its location by solving the following problem:

$$\left(\hat{\boldsymbol{x}}_i^{(t+1)}, \eta_i\right) = \arg\min_{\boldsymbol{x}_i, \eta_i} \sum_{j \in \mathcal{E}_{\mathcal{A}_i} \cup \mathcal{E}_{\mathcal{T}_i}} \left(\zeta_{ij}\|\boldsymbol{x}_i - \hat{\boldsymbol{a}}_j\| - \eta_i d_0\right)^2$$
$$+ \sum_{j \in \mathcal{E}_{\mathcal{A}_i}} \left(\boldsymbol{c}_{ij}^T(\boldsymbol{x}_i - \boldsymbol{a}_j)\right)^2 + \sum_{j \in \mathcal{E}_{\mathcal{A}_i}} \left(\boldsymbol{k}_{ij}^T(\boldsymbol{x}_i - \boldsymbol{a}_j) - \|\boldsymbol{x}_i - \boldsymbol{a}_j\| \cos(\alpha_{ij}^{\mathcal{A}})\right)^2. \tag{2.42}$$

By applying similar procedure as in the previous section, we obtain the following SOCP estimator:

$$\underset{\boldsymbol{x}_i, \eta_i, \boldsymbol{r}, \boldsymbol{z}, \boldsymbol{g}, \boldsymbol{p}, e_1, e_2, e_3}{\text{minimize}} \quad e_1 + e_2 + e_3$$

subject to

$$\begin{aligned} \|\boldsymbol{x}_i - \hat{\boldsymbol{a}}_j\| &\le r_{ij}, \ \forall (i,j) \in \mathcal{E}_{\mathcal{A}} \cup \mathcal{E}_{\mathcal{T}}, \\ z_{ij} &= \zeta_{ij} r_{ij} - \eta_i d_0, \ \forall (i,j) \in \mathcal{E}_{\mathcal{A}} \cup \mathcal{E}_{\mathcal{T}}, \\ g_{ij} &= \boldsymbol{c}_{ij}^T(\boldsymbol{x}_i - \boldsymbol{a}_j), \ \forall (i,j) \in \mathcal{E}_{\mathcal{A}}, \\ p_{ij} &= \boldsymbol{k}_{ij}^T(\boldsymbol{x}_i - \boldsymbol{a}_j) - r_{ij}\cos(\alpha_{ij}^{\mathcal{A}}), \ \forall (i,j) \in \mathcal{E}_{\mathcal{A}}, \\ \left\|\begin{bmatrix} 2\boldsymbol{z} \\ e_1 - 1 \end{bmatrix}\right\| &\le e_1 + 1, \left\|\begin{bmatrix} 2\boldsymbol{g} \\ e_2 - 1 \end{bmatrix}\right\| \le e_2 + 1, \left\|\begin{bmatrix} 2\boldsymbol{p} \\ e_3 - 1 \end{bmatrix}\right\| \le e_3 + 1. \end{aligned} \tag{2.43}$$

The problem in (2.43) is a classical SOCP, where the objective function and equality constraints are affine, and the inequality constraints are second-order cone constraints [56].

Algorithm 2 outlines the proposed SOCP algorithm for unknown P_i's. Lines 5–10 are performed concurrently by all targets $i \in \mathcal{C}_c$, which might reduce the running time of the algorithm. At Line 6, we solve (2.43) S number of times, after which we start calculating the ML estimate of P_{0i}, $\widehat{P}_{0i}$, and switch to solving (2.37) as if P_{0i} is known. Line 7 is introduced to avoid the oscillation in the location estimates. At Line 10, the location updates, $\hat{\boldsymbol{x}}_i^{(t+1)}$ $\forall i \in \mathcal{T}$, are broadcasted to neighbors of i. In the remaining text, we label Algorithm 2 as "uSOCP".

Algorithm 2 The proposed distributed uSOCP algorithm

Require: $\hat{\boldsymbol{x}}_i^{(0)}$, $\forall i \in \mathcal{T}$, $\boldsymbol{a}_j$, $\forall j \in \mathcal{A}$, $\mathcal{C}$, S, P_0^{Low}, P_0^{Up}, $T_{\max}$
1: **Initialize:** $t \leftarrow 0$
2: **repeat**
3:      **for** $c = 1, ..., \mathcal{C}$ **do**
4:          **for all** $i \in \mathcal{C}_c$ (in parallel) **do**
5:              Collect $\hat{\boldsymbol{a}}_j$, $\forall j \in \mathcal{E}_{\mathcal{A}_i} \cup \mathcal{E}_{\mathcal{T}_i}$
6:              $\hat{\boldsymbol{x}}_i^{(t+1)} \leftarrow \begin{cases} \text{solve (2.43)}, & \text{if } t < S, \\ \text{solve (2.37) using } \widehat{P}_{0i}, & \text{if } t \geq S \end{cases}$
7:              **if** $\frac{\|\hat{\boldsymbol{x}}_i^{(t+1)} - \hat{\boldsymbol{x}}_i^{(t)}\|}{\|\hat{\boldsymbol{x}}_i^{(t)}\|} > 1$ **then**
8:                  $\hat{\boldsymbol{x}}_i^{(t+1)} \leftarrow \hat{\boldsymbol{x}}_i^{(t)}$
9:              **end if**
10:             Broadcast $\hat{\boldsymbol{x}}_i^{(t+1)}$ to $\hat{\boldsymbol{a}}_j$, $\forall j \in \mathcal{E}_{\mathcal{A}_i} \cup \mathcal{E}_{\mathcal{T}_i}$
11:         **end for**
12:     **end for**
13:     $t \leftarrow t + 1$
14:     **if** $t > S$ **then**
15:         **for all** $i \in \mathcal{T}$ (in parallel) **do**
16:             $\widehat{P}_{0i} \leftarrow \frac{\sum_{j \in \mathcal{E}_{\mathcal{A}_i} \cup \mathcal{E}_{\mathcal{T}_i}} P_{ij} + 10\gamma \log_{10} \frac{\|\hat{\boldsymbol{x}}^{(t)} - \hat{\boldsymbol{a}}_j\|}{d_0}}{|\mathcal{E}_{\mathcal{A}_i}| + |\mathcal{E}_{\mathcal{T}_i}|}$
17:             **if** $\widehat{P}_{0i} < P_0^{\text{Low}}$ **then**
18:                 $\widehat{P}_{0i} \leftarrow P_0^{\text{Low}}$
19:             **else if** $\widehat{P}_{0i} > P_0^{\text{Up}}$ **then**
20:                 $\widehat{P}_{0i} \leftarrow P_0^{\text{Up}}$
21:             **end if**
22:         **end for**
23:     **end if**
24: **until** $t < T_{\max}$

2.3.4 Complexity Analysis

In order to evaluate the overall performance of a localization algorithm, it is necessary to analyze the trade off between the estimation accuracy and computational complexity. In this section, we investigate computational complexity of the considered algorithms. According to [72], the worst case computational complexity of an SOCP is:

$$\mathcal{O}\left(\sqrt{L}\left(m^2 \sum_{i=1}^{L} n_i + \sum_{i=1}^{L} n_i^2 + m^3\right)\right), \tag{2.44}$$

where L is the number of the second-order cone constraints, m is the number of the equality constraints, and n_i is the dimension of the i-th second-order cone.

Assuming that $N_{\max}$ is the maximum number of steps in the bisection procedure, Table 2.2 provides a summary of the worst case computational complexities of the considered algorithms. In Table 2.2, the labels "SDP" and "uSOCP2" are used to denote the centralized SDP algorithm in [36] and the distributed SOCP algorithm in [38], respectively, which will be used later on in Section 2.3.5 to offer a better understanding of the performance of the proposed algorithms.

Table 2.2 shows that the computational complexity of a distributed algorithm depends mainly on the size of neighborhood fragments, rather than the total number of sensors in a WSN. Theoretically, it is possible to have a fully connected network, *i.e.*, $|\mathcal{E}_{\mathcal{A}_i}| + |\mathcal{E}_{\mathcal{T}_i}| = M + N - 1, \forall i \in \mathcal{T}$. However, in practice, the size of the neighborhood fragments are much smaller, due to energy restrictions (limited R). Therefore, distributed algorithms are a preferable solution in large-scale and highly-dense networks, since adding more sensors in the network will not have a severe impact on the size of neighborhood fragments. Table 2.2 also reveals that the proposed distributed SOCP algorithms are computationally more demanding than the proposed SR-WLS one. This result is not surprising, since the SOCP approach employs sophisticated mathematical tools, whereas the SR-WLS approach applies a

Table 2.2 Computational complexity of the considered algorithms

Algorithm	Complexity
SOCP	$T_{\max} \times M \times \mathcal{O}\left(\left(\max_i \{3\lvert\mathcal{E}_{\mathcal{A}_i}\rvert + \lvert\mathcal{E}_{\mathcal{T}_i}\rvert\}\right)^{3.5}\right)$
SR-WLS	$T_{\max} \times M \times \mathcal{O}\left(N_{\max} \times \max_i \{3\lvert\mathcal{E}_{\mathcal{A}_i}\rvert + \lvert\mathcal{E}_{\mathcal{T}_i}\rvert\}\right)$
uSOCP	$T_{\max} \times M \times \mathcal{O}\left(\left(\max_i \{3\lvert\mathcal{E}_{\mathcal{A}_i}\rvert + \lvert\mathcal{E}_{\mathcal{T}_i}\rvert\}\right)^{3.5}\right)$
SDP	$\mathcal{O}\left(\sqrt{3M}\left(81M^4\left(N + \frac{M}{2}\right)^2\right)\right)$
uSOCP2	$T_{\max} \times M \times \mathcal{O}\Big(\max_i \Big\{ \sqrt{3\lvert\mathcal{E}_{\mathcal{A}_i}\rvert + \lvert\mathcal{E}_{\mathcal{T}_i}\rvert}\Big((3\lvert\mathcal{E}_{\mathcal{A}_i}\rvert)^2 (3\lvert\mathcal{E}_{\mathcal{A}_i}\rvert + \lvert\mathcal{E}_{\mathcal{T}_i}\rvert)\rvert + (3\lvert\mathcal{E}_{\mathcal{A}_i}\rvert + \lvert\mathcal{E}_{\mathcal{T}_i}\rvert)^2\Big)\Big\}\Big)$

bisection procedure to solve the localization problem. Nevertheless, higher complexity of the proposed SOCP algorithms is justified by their superior performance in terms of the estimation accuracy and convergence, as we will see in Section 2.3.5.

2.3.5 Performance Results

In this section, we present a set of results in order to asses the performance of the proposed approaches in terms of the estimation accuracy and convergence. All of the presented algorithms were solved by using the MATLAB package `CVX` [71], where the solver is SeDuMi [73]. In order to demonstrate the benefit of fusing two radio measurements versus traditional localization systems, we include also the performance results of the proposed methods when only RSS measurements are employed, called here "SOCP$_{\text{RSS}}$" and "SR-WLS$_{\text{RSS}}$". To provide a performance benchmark, we employ also the existing distributed SOCP approach for unknown P_i's [38] labelled as "uSOCP2", as well as the centralized cooperative approach described in [36] for known P_i's which is used as a lower bound on the performance of the distributed approaches, denoted as "SDP".

A random deployment of M targets and N anchors inside a cube region of length B in each Monte Carlo (M_c) run is considered. Random deployment of sensors is of particular interest, since the localization algorithms are tested against various network topologies in order to asses their robustness. In favor of making the comparison of the considered approaches as fair as possible, we first obtained $M_c = 500$ targets' and anchors' locations, as well as noise realizations between two sensors $\forall(i,j) \in \mathcal{E}_\mathcal{A} \cup \mathcal{E}_\mathcal{T}, i \neq j$, in each M_c run. Furthermore, we made sure that the network graph is connected in each M_c run. We then solved the localization problem with the considered approaches for those scenarios. In all simulations presented here, the reference distance was set to $d_0 = 1$ m, the communication range of a sensor to $R = 6.5$ m, the maximum number of steps in the bisection procedure to $N_{\max} = 30$ and the PLE was fixed to $\gamma = 3$. The true value of the reference power is drawn from a uniform distribution on an interval $[P_0^{\text{Low}}, P_0^{\text{Up}}]$, *i.e.*, $P_{0i} \in \mathcal{U}[P_0^{\text{Low}}, P_0^{\text{Up}}]$ dBm. Also, to account for a realistic measurement model mismatch and test the robustness of the new algorithms to imperfect knowledge of the PLE, the true PLE was drawn from $\gamma_{ij} \in \mathcal{U}[2.7, 3.3], \forall(i,j) \in \mathcal{E}_\mathcal{A} \cup \mathcal{E}_\mathcal{T}, i \neq j$. Finally, we assumed that the initial guess of the targets' locations, $\hat{\boldsymbol{X}}^{(0)}$, is in the intersection of the big diagonals of the cube area.

The performance metric is the NRMSE, defined as

$$\text{NRMSE} = \sqrt{\frac{1}{MM_c}\sum_{i=1}^{M_c}\sum_{j=1}^{M}\|\boldsymbol{x}_{ij} - \widehat{\boldsymbol{x}}_{ij}\|^2},$$

where $\widehat{\boldsymbol{x}}_{ij}$ denotes the estimate of the true location of the j-th target, $\boldsymbol{x}_{ij}$, in the i-th Monte Carlo run.

Figure 2.17 illustrates the NRMSE versus t performance of the considered approaches when $N = 20$ and $M = 50$. From Figure 2.17, we can see that the performance of all considered algorithms betters as t grows, as anticipated. Furthermore, it can be noticed that the "uSOCP" curve gets saturated at $t = 3$. Hence, at this point we start estimating P_{0i}'s, and continue our algorithm as if P_{0i}'s are known. This fact explains the sudden curve drop after $t = 3$. One can argue that the proposed "uSOCP" algorithm shows excellent performance, outperforming noticeably the existing "uSOCP2" approach and achieving the lower bound provided by its counterpart for known P_i's. Also, it can be seen that the proposed hybrid methods outperform considerably their traditional counterparts that utilize RSS measurements only. Moreover, the "SR-WLS" method performs better than the "SOCP_{RSS}" method in every iteration. This is important to note because the later method is computationally more demanding due to the use of sophisticated mathematical tools, which shows that even a simple algorithm such as the one based on bisection procedure can produce high estimation accuracy when two radio measurements are combined. One can perceive that all major changes in the performance for the considered algorithms take place in the first few iterations ($t \leq 10$ or $t \leq 20$), and that the performance gain is negligible afterwards. This result is very important because it shows that our approaches require a low number of signal transmissions, which might enhance the utilization efficiency of the radio spectrum, a precious resource for wireless communications. It also shows that our algorithms are energy efficient; the communication phase is much more expensive (in terms of energy) than the data processing one [1]. Finally, the proposed SOCP performs outstanding, very close to the lower bound provided by the centralized "SDP" approach in just a few iterations.

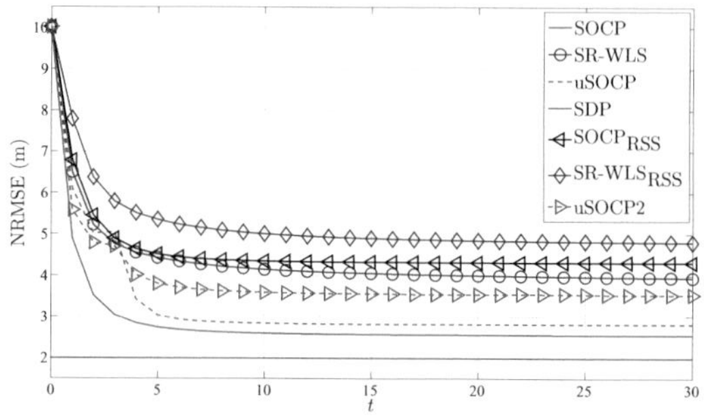

Figure 2.17 NRMSE versus t comparison, when $N = 20$, $M = 50$, $R = 6.5$ m, $\sigma_{n_{ij}} = 3$ dB, $\sigma_{m_{ij}} = 6$ deg, $\sigma_{v_{ij}} = 6$ deg, $\gamma_{ij} \in \mathcal{U}[2.7, 3.3]$, $\gamma = 3$, $B = 20$ m, $P_{0i} \in \mathcal{U}[-12, -8]$ dBm, $d_0 = 1$ m, $M_c = 500$.

Figure 2.18 illustrates the NRMSE versus t performance of the considered approaches when $N = 30$ and $M = 50$. Figures 2.17 and 2.18 reveal that the performance of all algorithms improves significantly as more anchors are added into the network. This behavior is expected, since when N grows more reliable information and more AoA measurements are available in the network. Furthermore, Figure 2.18 exhibits that the proposed hybrid algorithms outperform their RSS counterparts, and that they can be stopped after just 5–10 iterations. Finally, although the new methods were derived under the assumption

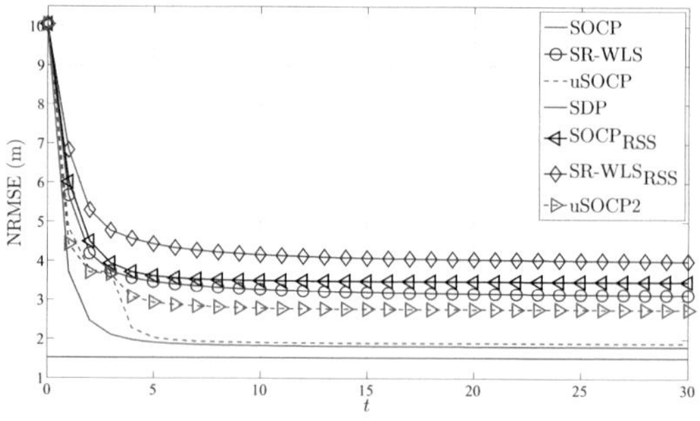

Figure 2.18 NRMSE versus t comparison, when $N = 30$, $M = 50$, $R = 6.5$ m, $\sigma_{n_{ij}} = 3$ dB, $\sigma_{m_{ij}} = 6$ deg, $\sigma_{v_{ij}} = 6$ deg, $\gamma_{ij} \in \mathcal{U}[2.7, 3.3]$, $\gamma = 3$, $B = 20$ m, $P_{0i} \in \mathcal{U}[-12, -8]$ dBm, $d_0 = 1$ m, $M_c = 500$.

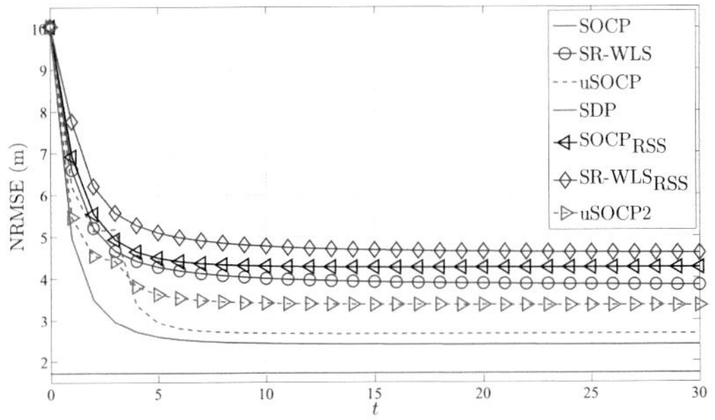

Figure 2.19 NRMSE versus t comparison, when $N = 20$, $M = 60$, $R = 6.5$ m, $\sigma_{n_{ij}} = 3$ dB, $\sigma_{m_{ij}} = 6$ deg, $\sigma_{v_{ij}} = 6$ deg, $\gamma_{ij} \in \mathcal{U}[2.7, 3.3]$, $\gamma = 3$, $B = 20$ m, $P_{0i} \in \mathcal{U}[-12, -8]$ dBm, $d_0 = 1$ m, $M_c = 500$.

that the noise is small, we can see that they work excellent even when the assumption does not hold.

Figure 2.19 illustrates the NRMSE versus t performance of the considered approaches when $N = 20$ and $M = 60$. From Figures 2.17 and 2.19 it can be seen that the distributed approaches require a slightly higher number of iterations to converge when M is increased. However, the estimation accuracy of the considered algorithms does not deteriorate when more targets are added in the network; it actually betters when M is increased. Finally, Figure 2.19 confirms the effectiveness of using the combined measurements in hybrid systems in comparison with using only a single measurement[3].

In Figures 2.20, 2.21 and 2.22 we investigate the impact of the quality of RSS and AoA measurements on the performance of the considered approaches. More precisely, Figures 2.20, 2.21 and 2.22 respectively illustrate the NRMSE versus $\sigma_{n_{ij}}$ (dB), $\sigma_{m_{ij}}$ (deg) and $\sigma_{v_{ij}}$ (deg) comparison, when $N = 20$, $M = 50$, $R = 6.5$ m, and $T_{\max} = 30$. In these figures, we can observe that the performance of all algorithms degrades as the quality of a certain measurement drops, as expected. It can also be seen that the quality of the RSS measurements has the most significant impact on the performance of the proposed algorithms, while the error in the azimuth and elevation angle measurements have

[3]Actually, in Figures 2.17, 2.18 and 2.19 we have performed the simulations with $T_{\max} = 200$ iterations in order to make sure that the considered approaches converge. In favour of a better overview, here, we present only the results for the first $t = 30$ iterations.

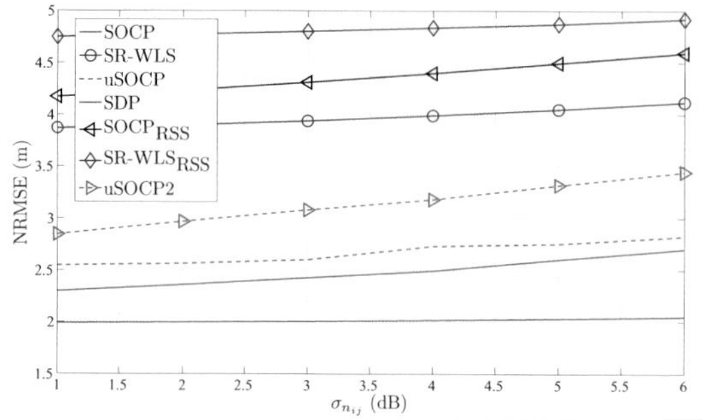

Figure 2.20 NRMSE versus $\sigma_{n_{ij}}$ (dB) comparison, when $N = 20$, $M = 50$, $R = 6.5$ m, $\sigma_{m_{ij}} = 1$ deg, $\sigma_{v_{ij}} = 1$ deg, $\gamma_{ij} \in \mathcal{U}[2.7, 3.3]$, $\gamma = 3$, $T_{\max} = 30$, $B = 20$ m, $P_{0i} \in \mathcal{U}[-12, -8]$ dBm, $d_0 = 1$ m, $M_c = 500$.

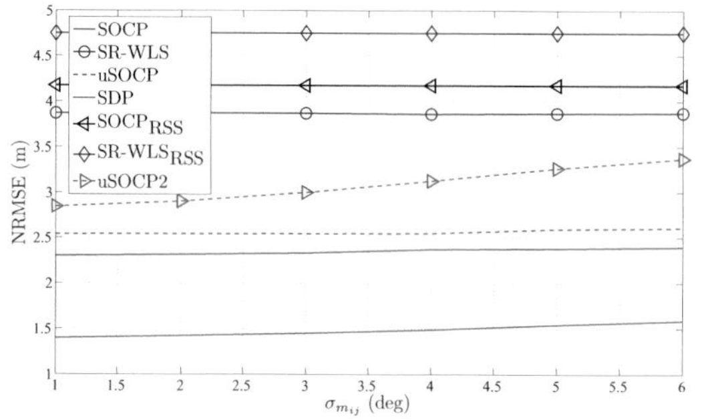

Figure 2.21 NRMSE versus $\sigma_{m_{ij}}$ (deg) comparison, when $N = 20$, $M = 50$, $R = 6.5$ m, $\sigma_{n_{ij}} = 1$ dB, $\sigma_{v_{ij}} = 1$ deg, $\gamma_{ij} \in \mathcal{U}[2.7, 3.3]$, $\gamma = 3$, $T_{\max} = 30$, $B = 20$ m, $P_{0i} \in \mathcal{U}[-12, -8]$ dBm, $d_0 = 1$ m, $M_c = 500$.

marginal influence on the performance. This is not surprising, since the error of a few degrees in AoA measurements does not impair considerably their quality on a fairly short distance (communication range of all sensors is restricted to $R = 6.5$ m), as shown in Figures 2.21 and 2.22. On the other hand, RSS measurements are notoriously unpredictable [1]. Nonetheless, we can see from Figure 2.20 that the performance loss is lower than 15 % for the "SOCP" and "uSOCP", and 10 % for the "SR-WLS", which is relatively low for the considered error span. Finally, from the figures, we can see that the proposed "uSOCP" outperforms the existing "uSOCP2" for all settings.

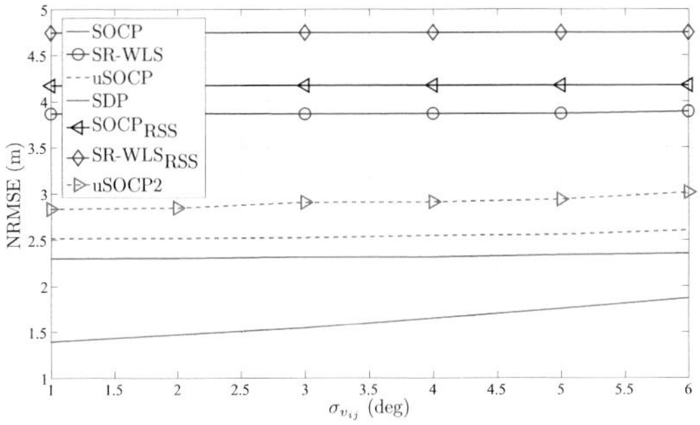

Figure 2.22 NRMSE versus $\sigma_{v_{ij}}$ (deg) comparison, when $N = 20$, $M = 50$, $R = 6.5$ m, $\sigma_{n_{ij}} = 1$ dB, $\sigma_{m_{ij}} = 1$ deg, $\gamma_{ij} \in \mathcal{U}[2.7, 3.3]$, $\gamma = 3$, $T_{\max} = 30$, $B = 20$ m, $P_{0i} \in \mathcal{U}[-12, -8]$ dBm, $d_0 = 1$ m, $M_c = 500$.

2.3.6 Conclusions

In this section, we proposed two novel distributed algorithms to solve the RSS/AoA localization problem for known transmit powers based on SOCP relaxation technique and GTRS framework. The proposed SOCP algorithm provides exceptional localization accuracy in just a few iterations. Our algorithm based on GTRS framework is solved via a simple bisection procedure, and it represents an excellent alternative to our SOCP algorithm, since its somewhat lower accuracy is compensated with linear computational complexity. We also show that the proposed SOCP algorithm for known transmit power can be generalized to the case where the transmit powers are different and not known. Our simulation results show that all of the proposed algorithms efficiently solve the very challenging cooperative localization problem, both in terms of the estimation accuracy and the convergence; the SOCP-based algorithm achieves the lower bound provided by the centralized SDP algorithm in only a few iterations, and outperforms notably the existing distributed approach. Furthermore, the simulation results confirmed the robustness of the proposed algorithms to the imperfect knowledge of the PLE, which is a very important practical scenario.

3

Target Tracking

3.1 Chapter Summary

This chapter addresses the problem of tracking of a moving target by using coupled RSS and AoA measurements. The remainder of the chapter is organized as follows.

Section 3.2 describes the existing work in the area of RSS-AoA-based target tracking problem, and highlights our contributions. In Section 3.3 we introduce the target state transition model as well as the measurement model, and we formulate the target tracking problem by using the Bayesian approach. Section 3.4 describes our technique used to *linearize* the measurement model. Section 3.5 present the derivation of our tracking algorithms, as well as our navigation routine for sensors' mobility management. In Section 3.6, simulation results are presented for two different target trajectories in order to validate the performance of our algorithms. Finally, Section 3.7 summarizes the main conclusions.

3.2 Introduction

The problem of accurate localization of a moving object in real-time has motivated a great deal of scientific research recently, owing to a constant growth of the range of enabling devices and technologies, and the requirement for seamless solutions in location-based services [16, 82–87]. In order to maintain low implementation costs, making use of existing technologies (such as terrestrial radio frequency sources) when providing a solution to the object tracking problem is strongly encouraged. These include for example, time of arrival, RSS, AoA, or a combination of them [20, 23, 36, 38, 57, 87–91].

3.2.1 Related Work

The authors in [20, 23, 36, 38, 57, 88, 89, 91] considered only the *classical* target localization problem, where they disregarded the prior knowledge and gave all importance to observations exclusively. The works in [83, 84, 87], investigated the target tracking problem, where the observations were combined with some prior knowledge to enhance the estimation accuracy. However, they all examined pure RSS-based target tracking problem only. In [86], the authos investigated the target tracking problem by employing hybrid, RSS and AoA, measurements. Both Kalman filter (KF) and particle filter (PF) were proposed in [86], as well as a generalized pattern search method for estimating the PLE for each link in every time step. Nevertheless, the authors considered the tracking problem with static anchors only. The works presented in [11, 92], tackled the target tracking problem with mobile sensors navigation. Still, hybrid RSS-AoA target tracking problem was not a part of their study.

3.2.2 Contribution

In this work, the problem of tracking a mobile target by employing hybrid RSS-AoA measurements is considered. We assume that the target transmit power is unknown, and start by describing our *linearization* process of the highly non-linear observation model. Next, we combine the prior knowledge given by state transition model with the *linearized* model in order to enhance the estimation accuracy. Then, is it shown that the application of the MAP and the KF criterion is straightforward, resulting in a novel MAP and a novel KF algorithm. Finally, we propose a simple navigation routine to manage sensors' mobility, which leads to great improvement in the estimation accuracy, even for lower number of sensors. A realistic scenario where the PLE and the true sensors' locations are not perfectly known is also taken into consideration.

3.3 Problem Formulation

We consider a WSN[1] composed of N mobile sensors with known locations, $\boldsymbol{a}_{i,t} = [a_{ix,t}, a_{iy,t}]^T$ for $i = 1, ..., N$, and a moving target whose

[1]For simplicity and without loss of generality, this work focuses on 2-dimensional scenario. The extension to 3-dimensional scenario is straightforward.

location, $\boldsymbol{x}_t = [x_{\mathrm{x},t}, x_{\mathrm{y},t}]^T$, we desire to determine at each time instant t. For simplicity, we assume a constant velocity target motion model (*e.g.*, perturbed only by wind gust) such that the velocity components in the x and y directions at time t are given by

$$\boldsymbol{v}_t = \boldsymbol{v}_{t-1} + \boldsymbol{r}_{v,t}, \tag{3.1}$$

where $\boldsymbol{r}_{v,t}$ represents the noise perturbations. Hence, from the equations of motion [46], the target location at time t is

$$\boldsymbol{x}_t = \boldsymbol{x}_{t-1} + \boldsymbol{v}_{t-1}\Delta + \boldsymbol{r}_{x,t}, \tag{3.2}$$

where Δ and $\boldsymbol{r}_{x,t}$ are the sampling interval between two consecutive time steps and location process noise, respectively. Now, if we describe the target state at t by its location and velocity, *i.e.*, $\boldsymbol{\theta}_t = [\boldsymbol{x}_t^T, \boldsymbol{v}_t^T]^T$, from (3.1) and (3.2) we get

$$\boldsymbol{\theta}_t = \boldsymbol{S}\,\boldsymbol{\theta}_{t-1} + \boldsymbol{r}_t, \tag{3.3}$$

where $\boldsymbol{r}_t = [\boldsymbol{r}_{x,t}^T, \boldsymbol{r}_{v,t}^T]^T$ is the state process noise [82–87], assumed to be zero-mean Gaussian with a covariance matrix $\boldsymbol{Q}$, *i.e.*, $\boldsymbol{r}_t \sim \mathcal{N}(\boldsymbol{0}, \boldsymbol{Q})$, where $\boldsymbol{Q}$ is defined as

$$\boldsymbol{Q} = q \begin{bmatrix} \frac{\Delta^3}{3} & 0 & \frac{\Delta^2}{2} & 0 \\ 0 & \frac{\Delta^3}{3} & 0 & \frac{\Delta^2}{2} \\ \frac{\Delta^2}{2} & 0 & \Delta & 0 \\ 0 & \frac{\Delta^2}{2} & 0 & \Delta \end{bmatrix},$$

with q denoting the state process noise intensity [82, 84, 93]. The symbol $\boldsymbol{S}$ in (3.3) stands for the state transition matrix, which models the state dynamics and is given by

$$\boldsymbol{S} = \begin{bmatrix} 1 & 0 & \Delta & 0 \\ 0 & 1 & 0 & \Delta \\ 0 & 0 & 1 & 0 \\ 0 & 0 & 0 & 1 \end{bmatrix}.$$

A detailed derivation of the state transition model, as well as the matrices $\boldsymbol{S}$ and $\boldsymbol{Q}$ is given in Appendix B.

At each time instant, the target emits a signal to sensors which withdraw the RSS and AoA information from it. Thus, the measurement equation can be formulated as

$$\boldsymbol{z}_t = \boldsymbol{h}(\boldsymbol{x}_t) + \boldsymbol{n}_t, \tag{3.4}$$

where $\boldsymbol{z}_t = [\boldsymbol{P}_t^T, \boldsymbol{\phi}_t^T]^T$ ($\boldsymbol{z}_t \in \mathbb{R}^{2N}$) is the observation vector comprising RSS, $\boldsymbol{P}_t = [P_{i,t}]^T$, and AoA, $\boldsymbol{\phi}_t = [\phi_{i,t}]^T$, measurements at time instant t. The function $\boldsymbol{h}(\boldsymbol{x}_t)$ in (3.4) is defined as $h_i(\boldsymbol{x}_t) = P_0 - 10\gamma \log_{10} \frac{\|\boldsymbol{x}_t - \boldsymbol{a}_{i,t}\|}{d_0}$ for $i = 1, ..., N$ [61], and $h_i(\boldsymbol{x}_t) = \tan^{-1}\left(\frac{x_{y,t} - a_{iy,t}}{x_{x,t} - a_{ix,t}}\right)$ for $i = N + 1, ..., 2N$ [30]. The measurement noise, $\boldsymbol{n}_t$, is modeled as $\boldsymbol{n}_t \sim \mathcal{N}(\boldsymbol{0}, \boldsymbol{C})$, where the noise covariance is defined as $\boldsymbol{C} = \text{diag}([\sigma_{n_i}^2, \sigma_{m_i}^2]) \otimes \boldsymbol{I}_4$, with σ_{n_i} (dB) and σ_{m_i} (rad) being the noise standard deviation of the RSS and AoA measurements, respectively, $\boldsymbol{I}_M$ denoting the identity matrix of size M and symbol $\otimes$ representing the Kronecker product.

In Bayesian estimation theory, the prior knowledge, obtained through the state transition model (3.3), is combined with the noisy observations (3.4) to obtain the marginal posterior PDF, $p(\boldsymbol{\theta}_t|\boldsymbol{z}_{1:t})$. Through $p(\boldsymbol{\theta}_t|\boldsymbol{z}_{1:t})$ we can quantify the belief we have in the values of the state $\boldsymbol{\theta}_t$ given all the past measurements $\boldsymbol{z}_{1:t}$ and obtain an estimate at any time instant we desire. The main steps of the Bayesian estimation are described below [82–87].

- *Initialization:* The marginal posterior PDF at $t = 0$ is set to the prior PDF $p(\boldsymbol{\theta}_0)$ of $\boldsymbol{\theta}_0$.
- *Prediction:* By using the state transition model (3.3), the predictive PDF of the state at t is given by

$$p(\boldsymbol{\theta}_t|\boldsymbol{z}_{1:t-1}) = \int p(\boldsymbol{\theta}_t|\boldsymbol{\theta}_{t-1}) p(\boldsymbol{\theta}_{t-1}|\boldsymbol{z}_{1:t-1}) d\boldsymbol{\theta}_{t-1}. \tag{3.5}$$

- *Update:* By following the Bayes' rule [82, 93], we have

$$p(\boldsymbol{\theta}_t|\boldsymbol{z}_{1:t}) = \frac{p(\boldsymbol{z}_t|\boldsymbol{\theta}_t) p(\boldsymbol{\theta}_t|\boldsymbol{z}_{1:t-1})}{p(\boldsymbol{z}_t|\boldsymbol{z}_{1:t-1})} \tag{3.6}$$

where $p(\boldsymbol{z}_t|\boldsymbol{\theta}_t)$ is the likelihood and $p(\boldsymbol{z}_t|\boldsymbol{z}_{1:t-1}) = \int p(\boldsymbol{z}_t|\boldsymbol{\theta}_t) p(\boldsymbol{\theta}_t|\boldsymbol{z}_{1:t-1}) d\boldsymbol{\theta}_t$ is just a normalizing constant, independent of $\boldsymbol{\theta}_t$, needed to insure that $p(\boldsymbol{\theta}_t|\boldsymbol{z}_{1:t})$ integrates to 1 [46]. In general, the marginal PDF at $t - 1$ cannot be calculated analytically, and the integral in (3.5) cannot be obtained analytically if the state model is non-linear. Therefore, some approximations are required in order to obtain $p(\boldsymbol{\theta}_t|\boldsymbol{z}_{1:t})$.

3.4 Linearization of the Measurement Model

In practice, network testing and calibration are often not the priority, especially in low-cost systems such as RSS [16]. Hence, some parameters, such as target transmit power, might not be calibrated, *i.e.*, not known beforehand. Not knowing the transmit power matches not knowing P_0 in (3.4) [16]. Therefore, in this section we will show how to *linearize* the measurement model for the case of unknown P_0.

Under the assumption that the noise power is small, by using Taylor series approximation from (3.4) we can write for the RSS model

$$\rho + \epsilon_i = \mu_{i,t} \|\boldsymbol{x}_t - \boldsymbol{a}_{i,t}\|, \text{ for } i = 1, ..., N, \tag{3.7}$$

where $\rho = \exp\left(\frac{P_0}{\eta\gamma}\right)$, $\eta = \frac{10}{\ln(10)}$, $\mu_{i,t} = \exp\left(\frac{P_{i,t}}{\eta\gamma}\right)$, and $\epsilon_i \sim \mathcal{N}(0, (\frac{\rho}{\eta\gamma}\sigma_{n_i})^2)$. By rearranging the terms and squaring (3.7), we get

$$\epsilon_i \approx \frac{\mu_{i,t}}{2} \|\boldsymbol{x}_t - \boldsymbol{a}_{i,t}\| - \frac{\rho}{2}, \tag{3.8}$$

where we disregarded the second-order noise terms. By converting from Cartesian to polar coordinates, we can express $\boldsymbol{x}_t - \boldsymbol{a}_{i,t} = r_{i,t}\boldsymbol{u}_{i,t}$: $r_{i,t} \geq 0, \|\boldsymbol{u}_{i,t}\| = 1$, where the unit vector can be obtained by employing the available AoA information, *i.e.*, $\boldsymbol{u}_i = [\cos(\phi_{i,t}), \sin(\phi_{i,t})]^T$. If we apply this conversion in (3.8) and multiply by $\boldsymbol{u}_{i,t}^T\boldsymbol{u}_{i,t}$ we get

$$\epsilon_i \approx \frac{\mu_{i,t}}{2} \boldsymbol{u}_{i,t}^T(\boldsymbol{x}_t - \boldsymbol{a}_{i,t}) - \frac{\rho}{2}. \tag{3.9}$$

Similarly, for the AoA model we can write from (3.4)

$$\boldsymbol{c}_{i,t}^T(\boldsymbol{x}_t - \boldsymbol{a}_{i,t}) \approx 0, \text{ for } i = N+1, ..., 2N, \tag{3.10}$$

where $\boldsymbol{c}_{i,t} = [-\sin(\phi_{i,t}), \cos(\phi_{i,t})]^T$.

Assuming that the noise term is sufficiently small and introducing weights, $\boldsymbol{w}_t = [\sqrt{w_{i,t}}]$, where $w_{i,t} = P_{i,t}/\sum_{i=1}^N P_{i,t}$, in (3.9) and (3.10) such that more importance is given to *nearby* links, gives

$$w_{i,t}\frac{\mu_{i,t}}{2} \boldsymbol{u}_{i,t}^T(\boldsymbol{x}_t - \boldsymbol{a}_{i,t}) \approx w_{i,t}\frac{\rho}{2}, \text{ for } i = 1, ..., N, \tag{3.11a}$$

$$w_{i,t}\boldsymbol{c}_{i,t}^T(\boldsymbol{x}_t - \boldsymbol{a}_{i,t}) \approx 0, \text{ for } i = N+1, ..., 2N. \tag{3.11b}$$

We can rewrite (3.11) in a linear vector form as

$$\boldsymbol{A}_t \boldsymbol{y}_t = \boldsymbol{b}_t, \tag{3.12}$$

where $\boldsymbol{y}_t = [\boldsymbol{x}_t^T, \rho]^T$, and

$$\boldsymbol{A}_t = \begin{bmatrix} \vdots & \vdots \\ w_{i,t}\frac{\mu_{i,t}}{2}\boldsymbol{u}_{i,t}^T & -\frac{w_{i,t}}{2} \\ \vdots & \vdots \\ w_{i,t}\boldsymbol{c}_i^T & 0 \\ \vdots & \vdots \end{bmatrix}, \ \boldsymbol{b}_t = \begin{bmatrix} \vdots \\ w_{i,t}\frac{\mu_{i,t}}{2}\boldsymbol{u}_{i,t}^T\boldsymbol{a}_{i,t} \\ \vdots \\ w_{i,t}\boldsymbol{c}_{i,t}^T\boldsymbol{a}_{i,t} \\ \vdots \end{bmatrix}.$$

By applying the LS criterion to the *linearized* measurement model in (3.12) we get

$$\hat{\boldsymbol{y}}_t = \underset{\boldsymbol{y}_t=[\boldsymbol{x}_t^T,\rho]^T}{\arg\min} \ \|\boldsymbol{A}_t\boldsymbol{y}_t - \boldsymbol{b}_t\|^2, \tag{3.13}$$

whose solution is readily obtained as $\hat{\boldsymbol{y}}_t = (\boldsymbol{A}_t^T\boldsymbol{A}_t)^{-1}(\boldsymbol{A}_t^T\boldsymbol{b}_t)$.

Subsequently, one could take advantage of the solution obtained through (3.13) to additionally improve its quality, *i.e.*, it could be exploited to find the ML estimate[2] of P_0, $\widehat{P}_0$, as

$$\widehat{P}_0 = \frac{\sum_{i=1}^{N} P_{i,t} + 10\gamma\log_{10}\frac{\|\hat{\boldsymbol{x}}_t - \boldsymbol{a}_{i,t}\|}{d_0}}{N} \tag{3.14}$$

and *linearize* the measurement model in a similar manner as before. First, calculate $\widehat{\rho} = \exp\left(\frac{\widehat{P}_0}{\eta\gamma}\right)$. Define weights $\widetilde{\boldsymbol{w}}_t = \sqrt{\widetilde{w}_{i,t}}$, where $\widetilde{w}_{i,t} = 1 - \frac{\widehat{d}_{i,t}}{\sum_{i=1}^{N}\widehat{d}_{i,t}}$ with $\widehat{d}_{i,t} = \exp\left(\frac{\widehat{P}_0 - P_{i,t}}{\eta\gamma}\right)$ being the ML estimate of the distance between the target and i-th anchor at time t. Then, by following similar steps as above, we can rewrite (3.9) and (3.10) in a vector form as

$$\widetilde{\boldsymbol{A}}_t\boldsymbol{x}_t = \widetilde{\boldsymbol{b}}_t, \tag{3.15}$$

[2]In the case where the true value of the transmit power is available beforehand, one would simply substitute $\widehat{P}_0$ by P_0 in the upcoming steps.

where

$$\widetilde{\boldsymbol{A}}_t = \begin{bmatrix} \vdots \\ \widetilde{w}_{i,t} \frac{\mu_{i,t}}{2} \boldsymbol{u}_{i,t}^T \\ \vdots \\ \widetilde{w}_{i,t} \boldsymbol{c}_i^T \\ \vdots \end{bmatrix}, \ \widetilde{\boldsymbol{b}}_t = \begin{bmatrix} \vdots \\ \widetilde{w}_{i,t} (\frac{\mu_{i,t}}{2} \boldsymbol{u}_{i,t}^T \boldsymbol{a}_{i,t} + \frac{\widehat{\rho}}{2}) \\ \vdots \\ w_{i,t} \boldsymbol{c}_{i,t}^T \boldsymbol{a}_{i,t} \\ \vdots \end{bmatrix}.$$

Hence, the localization problem can be posed in an LS form

$$\hat{\boldsymbol{x}}_t = \arg\min_{\boldsymbol{x}_t} \|\widetilde{\boldsymbol{A}}_t \boldsymbol{x}_t - \widetilde{\boldsymbol{b}}_t\|^2, \tag{3.16}$$

whose solution is obtained as $\hat{\boldsymbol{x}}_t = (\widetilde{\boldsymbol{A}}_t^T \widetilde{\boldsymbol{A}}_t)^{-1} (\widetilde{\boldsymbol{A}}_t^T \widetilde{\boldsymbol{b}}_t)$.

3.5 Target Tracking

3.5.1 Maximum A Posteriori Estimator

Within the Bayesian methodology, one of the most common criteria for determining a state estimate is the MAP criteria [46]. According to this estimation approach, we choose a state estimate, $\hat{\boldsymbol{\theta}}_{t|t}$, that maximizes the marginal PDF, *i.e.*,

$$\hat{\boldsymbol{\theta}}_{t|t} = \arg\max_{\boldsymbol{\theta}_t} p(\boldsymbol{\theta}_t|\boldsymbol{z}_{1:t}). \tag{3.17}$$

Based on (3.6), we observe that (3.17) is equivalent to maximization of $p(\boldsymbol{z}_t|\boldsymbol{\theta}_t)p(\boldsymbol{\theta}_t|\boldsymbol{z}_{1:t-1})$. This is evocative of the ML estimator except for the presence of the prior PDF. Consequently, the MAP estimator is

$$\begin{aligned} \hat{\boldsymbol{\theta}}_{t|t} &= \arg\max_{\boldsymbol{\theta}_t} p(\boldsymbol{z}_t|\boldsymbol{\theta}_t)p(\boldsymbol{\theta}_t|\boldsymbol{z}_{1:t-1}) \\ &= \arg\max_{\boldsymbol{\theta}_t} [\ln p(\boldsymbol{z}_t|\boldsymbol{\theta}_t) + \ln p(\boldsymbol{\theta}_t|\boldsymbol{z}_{1:t-1})]. \end{aligned} \tag{3.18}$$

The problem in (3.18) is highly non-convex and its analytical solution cannot be obtained in general. As such, some approximations are required in order to obtain $\hat{\boldsymbol{\theta}}_{t|t}$.

First, we can approximate $p(\boldsymbol{\theta}_{t-1}|\boldsymbol{z}_{1:t-1})$ as a Gaussian distribution [93], *i.e.*, $p(\boldsymbol{\theta}_{t-1}|\boldsymbol{z}_{1:t-1}) \sim \mathcal{N}(\hat{\boldsymbol{\theta}}_{t-1|t-1}, \hat{\boldsymbol{\Sigma}}_{t-1|t-1})$. Then, according to (3.5) we get

$$p(\boldsymbol{\theta}_t|\boldsymbol{z}_{1:t-1}) \approx \frac{1}{k_1} \exp\left(-\frac{1}{2}(\boldsymbol{\theta}_t - \hat{\boldsymbol{\theta}}_{t|t-1})^T \hat{\boldsymbol{\Sigma}}_{t|t-1}^{-1} (\boldsymbol{\theta}_t - \hat{\boldsymbol{\theta}}_{t|t-1})\right), \tag{3.19}$$

where k_1 is a constant, and $\hat{\boldsymbol{\theta}}_{t|t-1}$ and $\hat{\boldsymbol{\Sigma}}_{t|t-1}$ are the mean and the covariance of the one-step predicted state acquired through (3.3) as

$$\hat{\boldsymbol{\theta}}_{t|t-1} = \boldsymbol{S}\,\hat{\boldsymbol{\theta}}_{t-1|t-1} \tag{3.20a}$$

$$\hat{\boldsymbol{\Sigma}}_{t|t-1} = \boldsymbol{S}\,\hat{\boldsymbol{\Sigma}}_{t-1|t-1}\,\boldsymbol{S}^T + \boldsymbol{Q}. \tag{3.20b}$$

The likelihood function can be written as

$$p(\boldsymbol{z}_t|\boldsymbol{\theta}_t) = \frac{1}{k_2}\exp\left(-\frac{1}{2}\left(\boldsymbol{z}_t - \boldsymbol{h}(\boldsymbol{x}_t)\right)^T \boldsymbol{C}^{-1}\left(\boldsymbol{z}_t - \boldsymbol{h}(\boldsymbol{x}_t)\right)\right), \tag{3.21}$$

where k_2 is a constant. Then, according to (3.18) we have

$$\begin{aligned}\hat{\boldsymbol{\theta}}_{t|t} = \underset{\boldsymbol{\theta}_t}{\arg\min}\ &\left(\boldsymbol{z}_t - \boldsymbol{h}(\boldsymbol{x}_t)\right)^T \boldsymbol{C}^{-1}\left(\boldsymbol{z}_t - \boldsymbol{h}(\boldsymbol{x}_t)\right)\\ &+(\boldsymbol{\theta}_t - \hat{\boldsymbol{\theta}}_{t|t-1})^T\hat{\boldsymbol{\Sigma}}_{t|t-1}^{-1}(\boldsymbol{\theta}_t - \hat{\boldsymbol{\theta}}_{t|t-1}).\end{aligned} \tag{3.22}$$

We have shown in Section 3.4 how to tightly approximate the likelihood function. By following similar reasoning, we can rewrite (3.22) as

$$\hat{\boldsymbol{\theta}}_{t|t} = \underset{\boldsymbol{\theta}_t}{\arg\min}\ \|\boldsymbol{H}_t\boldsymbol{\theta}_t - \boldsymbol{f}_t\|^2, \tag{3.23}$$

where $\boldsymbol{H}_t = \left[\widetilde{\boldsymbol{A}}_t,\ \boldsymbol{0}_{2N\times 2};\ \hat{\boldsymbol{\Sigma}}_{t|t-1}^{-1/2}\right]$ ($\boldsymbol{H}_t \in \mathbb{R}^{(2N+4)\times 4}$), $\boldsymbol{f}_t = \left[\widetilde{\boldsymbol{b}}_t;\ \hat{\boldsymbol{\Sigma}}_{t|t-1}^{-1/2}\,\hat{\boldsymbol{\theta}}_{t|t-1}\right]$ ($\boldsymbol{f}_t \in \mathbb{R}^{2N+4}$), and $\boldsymbol{0}_{D\times L}$ is a D by L matrix of all zeros. The solution of (3.23) is obtained as $\hat{\boldsymbol{\theta}}_{t|t} = (\boldsymbol{H}_t^T\boldsymbol{H}_t)^{-1}(\boldsymbol{H}_t^T\boldsymbol{f}_t)$.

The step by step proposed MAP-based algorithm[3] for the case where the target transmit power is not known (labelled here as "uMAP") is outlined in Algorithm 3.

3.5.2 Kalman Filter

The KF may be thought of as a generalized sequential minimum mean square estimator of a signal embedded in noise, where the unknown

[3]Notice that in Algorithm 3 we do not update the state covariance matrix. Although this update could be accomplished through $\hat{\boldsymbol{\theta}}_{t|t}$ and the use of Karush-Kuhn-Tucker optimality conditions together with certain approximations (*e.g.* see the approach in [93]), it does not bring any gain to our uMAP algorithm, and we do not apply it here.

Algorithm 3 uMAP algorithm description

Require: $\boldsymbol{z}_t$, for $t = 0, ..., T-1$, $\boldsymbol{Q}$, $\boldsymbol{S}$

1: **Initialization:** $\hat{\boldsymbol{x}}_{0|0} \leftarrow$ (3.13), $\widehat{P}_0 \leftarrow$ (3.14), $\hat{\boldsymbol{\theta}}_{0|0} \leftarrow [\hat{\boldsymbol{x}}_{0|0}^T, 0, 0]^T$, $\hat{\boldsymbol{\Sigma}}_{t|t} \leftarrow \boldsymbol{I}_4$ for $t = 0, ..., T-1$

2: **for** $t = 1, ..., T-1$ **do**

3: **Prediction:**

4: – $\hat{\boldsymbol{\theta}}_{t|t-1} \leftarrow$ (3.20a)
– $\hat{\boldsymbol{\Sigma}}_{t|t-1} \leftarrow$ (3.20b)

5: **Update:**

6: – $\hat{\boldsymbol{\theta}}_{t|t} \leftarrow$ (3.23)
– $\widehat{P}_0 \leftarrow$ (3.14)

7: **end for**

parameters are allowed to evolve in time according a given dynamical model [46]. If the state and the measurement models are linear and the noise is assumed to be zero-mean with finite covariance, the KF provides the optimal solution in the LS sense [82].

Even though the measurement model (3.4) is non-linear, we can *linearize* it as in (3.15). Therefore, by following the KF recipe [46], the mean and the covariance are updated as

$$\hat{\boldsymbol{\theta}}_{t|t} = \hat{\boldsymbol{\theta}}_{t|t-1} + \boldsymbol{K}_t(\widetilde{\boldsymbol{b}}_t - \boldsymbol{G}_t\hat{\boldsymbol{\theta}}_{t|t-1}), \tag{3.24a}$$

$$\hat{\boldsymbol{\Sigma}}_{t|t} = (\boldsymbol{I}_4 - \boldsymbol{K}_t\boldsymbol{G}_t)\,\hat{\boldsymbol{\Sigma}}_{t|t-1}, \tag{3.24b}$$

where $\boldsymbol{K}_t$ is the Kalman gain at time instant t, and $\boldsymbol{G}_t = [\widetilde{\boldsymbol{A}}_t, \boldsymbol{0}_{2N\times 2}]$ ($\boldsymbol{G}_t \in \mathbb{R}^{2N\times 4}$).

The step by step proposed KF algorithm for the case where the target transmit power is not known (denoted here as "uKF") is outlined in Algorithm 4.

3.5.3 Sensor Navigation

Although the proposed algorithms described in Algorithm 3 and Algorithm 4 provide efficient solution to the target tracking problem, their estimation accuracy can be further enhanced. Until now, we have considered the sensors to be static, and only the target to be mobile. By allowing sensor mobility, such that they are permited to move in certain directions based on pre-established rules, we cannot only improve the estimation accuracy of the proposed algorithms, but do so with a reduced number of sensors. The price to pay for applying

Algorithm 4 uKF algorithm description

Require: $\boldsymbol{z}_t$, for $t = 0, ..., T-1$, $\boldsymbol{Q}$, $\boldsymbol{C}$, $\boldsymbol{S}$

1: **Initialization:** $\hat{\boldsymbol{x}}_{0|0} \leftarrow$ (3.13), $\widehat{P}_0 \leftarrow$ (3.14), $\hat{\boldsymbol{\theta}}_{0|0} \leftarrow [\hat{\boldsymbol{x}}_{0|0}^T, 0, 0]^T$, $\hat{\boldsymbol{\Sigma}}_{0|0} \leftarrow \boldsymbol{I}_4$

2: **for** $t = 1, ..., T-1$ **do**

3: **Prediction:**

4: – $\hat{\boldsymbol{\theta}}_{t|t-1} \leftarrow$ (3.20a)
– $\hat{\boldsymbol{\Sigma}}_{t|t-1} \leftarrow$ (3.20b)

5: $\boldsymbol{K}_t \leftarrow \hat{\boldsymbol{\Sigma}}_{t|t-1}\boldsymbol{G}_t^T(\boldsymbol{G}_t\hat{\boldsymbol{\Sigma}}_{t|t-1}\boldsymbol{G}_t^T + \boldsymbol{C})^{-1}$

6: **Update:**

7: – $\hat{\boldsymbol{\theta}}_{t|t} \leftarrow$ (3.24a)
– $\hat{\boldsymbol{\Sigma}}_{t|t} \leftarrow$ (3.24b)
– $\widehat{P}_0 \leftarrow$ (3.14)

8: **end for**

such a routine is somewhat increased computational cost (required for determining the direction of sensor movement) and increased energy consumption (depleted in the process of the actual sensor movement), in comparison with the static sensors routine. Nevertheless, the interest for target tracking problem using navigated mobile sensors is growing rapidly, especially in areas such as autonomous surveillance, automated data collection and monitoring to name a few [11, 92].

The proposed routine for sensor navigation is described in Algorithm 5. It represents a universal addition to the proposed uMAP and uKF algorithms, which is realized by incorporating lines 3–13 after line 6 in the uMAP, and after line 7 in the uKF. The basic idea of our navigation routine is to let the mobile sensors approach the target with the shortest possible path determined by the available information such as their estimated and the target's estimated location) at time instant t, until a certain threshold distance, τ. After the mobile sensors penetrate τ, our idea is to spread them around the target, so that we prevent possible sensor collision. More specifically, at line 4, each mobile sensor estimates its candidate location, $\breve{\boldsymbol{a}}_{i,t}$, by resorting only to its previous estimated location, the already available AoA measurement and its velocity, v_a. With this candidate location, the mobile sensors then estimate the possible distance from the estimated target location (as if they moved to the candidate location). If this estimated distance is not less than τ, the candidate location is accepted as the new estimated location of the mobile sensor, $\hat{\boldsymbol{a}}_{i,t}$, line 6; otherwise, the mobile sensors are spread around the target. To this end, we modify the angle measurement, $\breve{\phi}_{i,t-1}$ at line 9, and use this modified value to estimate the updated location of the mobile sensors, line 10. However, to account for a realistic model mismatches, at lines 7 and 11, we include noise perturbations within the sensors' actual movements, which result in imperfect knowledge about the mobile sensors' locations.

Algorithm 5 Sensor navigation algorithm description

Require: $\boldsymbol{a}_{i,0}$, $\phi_{i,t}$, for $i = 1, ..., N$, $t = 0, ..., T-1$, v_a, τ
1: **Initialization:** $\hat{\boldsymbol{a}}_{i,0} \leftarrow \boldsymbol{a}_{i,0}$
2: **for** $t = 1, ..., T-1$ **do**
3: **for** $i = 1, ..., N$ **do**
4: $\breve{\boldsymbol{a}}_{i,t} \leftarrow \hat{\boldsymbol{a}}_{i,t-1} + v_a\Delta[\cos(\phi_{i,t-1}), \sin(\phi_{i,t-1})]^T$
5: **if** $\|\breve{\boldsymbol{a}}_{i,t} - \hat{\boldsymbol{x}}_t\| \geq \tau$ **then**
6: $\hat{\boldsymbol{a}}_{i,t} \leftarrow \breve{\boldsymbol{a}}_{i,t}$
7: $\boldsymbol{a}_{i,t} \leftarrow \boldsymbol{a}_{i,t-1} + v_a\Delta[\cos(\phi_{i,t-1}), \sin(\phi_{i,t-1})]^T + \boldsymbol{r}_{x,t}$
8: **else**
9: $\breve{\phi}_{i,t} \leftarrow \phi_{i,t} + (-1)^i\,\pi/4$
10: $\hat{\boldsymbol{a}}_{i,t} \leftarrow \hat{\boldsymbol{a}}_{i,t-1} + v_a\Delta[\cos(\breve{\phi}_{i,t-1}), \sin(\breve{\phi}_{i,t-1})]^T$
11: $\boldsymbol{a}_{i,t} \leftarrow \boldsymbol{a}_{i,t-1} + v_a\Delta[\cos(\breve{\phi}_{i,t-1}), \sin(\breve{\phi}_{i,t-1})]^T + \boldsymbol{r}_{x,t}$
12: **end if**
13: **end for**
14: **end for**

3.6 Performance Results

3.6.1 Simulation Results

In this section, we validate the performance of the proposed algorithms through computer simulations. All of the presented algorithms were solved by using MATLAB. We consider two essentially different scenarios: one in which the target takes sharp manoeuvres and another one in which the target trajectory is more smooth; see Figure 3.1. The initial target location is at $[21, 20]^T$ (Fig,3.1a) and $[40, 15]^T$ (3.1b). The target state changes according to the state transition model (3.3), and at each time instant the radio measurements are generated in concordance with (3.4). Unless stated otherwise, the reference power is set to $P_0 = -10$ dBm, the initial location of $N = 3$ sensors is fixed at $\boldsymbol{a}_{i,0} = [[70, 10]^T, [40, 70]^T[10, 40]^T]$ and the PLE is set to $\gamma = 3$. However, to account for a more realistic measurement model mismatch, the true value of the PLE for each link was drawn from a uniform distribution on an interval $[2.7, 3.3]$, *i.e.*, $\gamma_{i,t} \sim \mathcal{U}[2.7, 3.3]$, at every

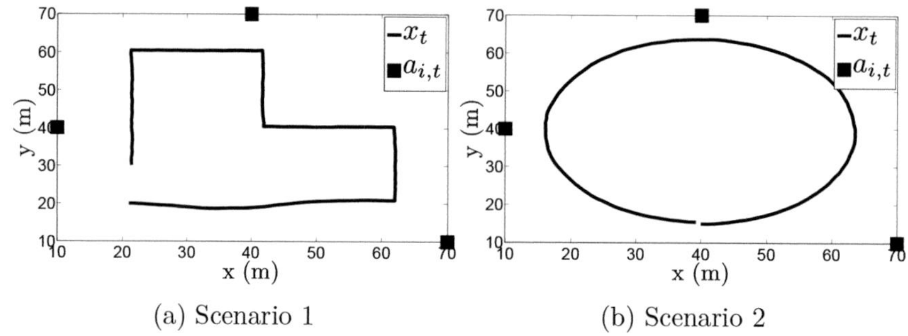

Figure 3.1 True target trajectory and mobile sensors' initial locations.

time instant. A sample is taken every $\Delta = 1$ s during $T = 150$ s trajectory duration in each Monte Carlo, $M_c = 1000$, run. In the case where sensor mobility is allowed, we set the threshold distance[4] to $\tau = 5$ m. Furthermore, $\sigma_{n_i} = 9$ dB, $\sigma_{m_i} = 4\pi/180$ rad, and $q = 2.5 \times 10^{-3}$ m^2/s^3. The performance metric used here is the RMSE, defined as $\text{RMSE}_t = \sqrt{\sum_{i=1}^{M_c} \frac{\|\boldsymbol{x}_{i,t} - \widehat{\boldsymbol{x}}_{i,t}\|^2}{M_c}}$, where $\widehat{\boldsymbol{x}}_{i,t}$ denotes the estimate of the true target location, $\boldsymbol{x}_{i,t}$, in the i-th M_c run at time instant t.

The performance of the proposed uMAP and uKF algorithms is compared with the existing KF in [86], where the initial target state was obtained by solving the LS method used in [86] to linearize the observation model[5]. Moreover, in favour of testing the belief that the Bayesian approaches (which integrate the prior knowledge with observations) outperform the *classical* ones (which disregard the prior knowledge and are based merely on observations), we show here the results for the sequential localization method in (3.16) with perfect knowledge about the target transmit power and PLE, denoted here by "WLS". Finally, to offer a lower bound on the performance of the proposed algorithms their counterparts for known target transmit power are also included, labelled here as "MAP" and "KF".

---

[4]In our simulations, we have also studied the influence of this parameter on the performance of the proposed algorithms. It was concluded that it has no significant impact on the performance, since similar results were attained for different values of τ (*e.g.*, $\tau = 10$ m or $\tau = 0$ m). However, we chose this particular value because it seems a reasonable practical threshold, keeping in mind the estimation error and noise influence to prevent sensor collision.

[5]Since the authors in [86] proposed also a method for PLE estimation, in order to make the comparison fair, the true value of the PLE for every link is considered perfectly known for the KF [86] at any time step.

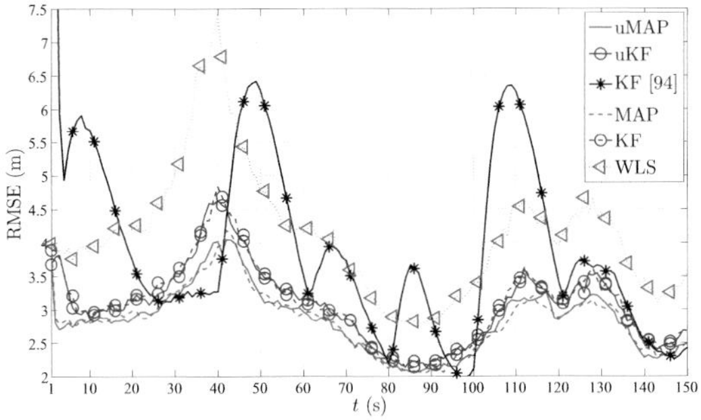

Figure 3.2 RMSE (m) versus t (s) comparison in the first scenario, when $N = 3$, $v_a = 0$ m/s, $\sigma_{n_i} = 9$ dB, $\sigma_{m_i} = 4\frac{\pi}{180}$ rad, $\gamma = 3$, $\gamma_i \sim \mathcal{U}[2.7, 3.3]$, $P_0 = -10$ dBm, $q = 2.5 \times 10^{-3}$ m^2/s^3, $M_c = 1000$.

Figure 3.2 illustrates the RMSE (m) versus t (s) comparison of all considered approaches in the first scenario, for the static sensors case, *i.e.*, $v_a = 0$ m/s. From it, we can observe that all algorithms suffer deteriorations at each sharp manoeuvre of the target, especially in the proximity of the sensors. This is somewhat anticipated, since the role of the prior knowledge is cancelled out with each sharp manoeuvre, and the vicinity of the target and any of the sensors creates a disbalance between the significance of that particular measurement and all of other ones. Nonetheless, all algorithms recover fairly quickly from these impairments. Furthermore, the figure shows that the proposed algorithms outperform the existing KF in [86] in general, as well as the *classical* approach for all t. Moreover, it is worth mentioning that our algorithms show robustness to not knowing the target transmit power, since they achieve their lower bounds given by their equivalents for known transmit power. Finally, the new algorithms behave excellent even for the case where the PLE is not perfectly known.

Figure 3.3 illustrates the RMSE (m) versus t (s) comparison of all considered approaches in the second scenario, for the static sensors case. From the figure, one can observe that the performance of all considered algorithms is significantly smoother in comparison with the first scenario. This behaviour is not surprising, since the target, although constantly changing its direction, is moving in a much smoother manner now. Figure 3.3 exhibits also superior performance of the proposed algorithms in general, and robustness to not knowing the transmit power.

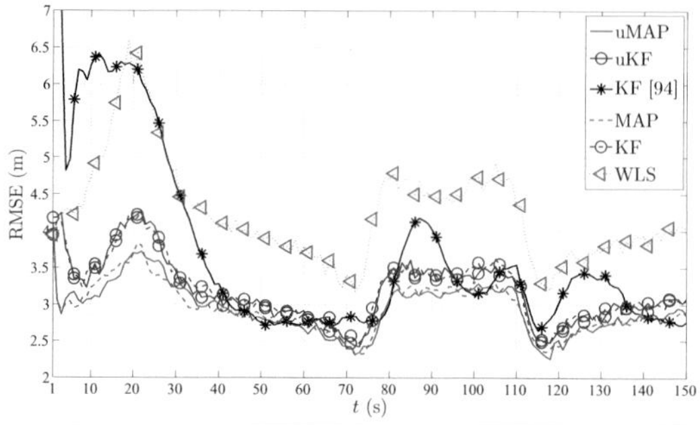

Figure 3.3 RMSE (m) versus t (s) comparison in the second scenario, when $N = 3$, $v_a = 0$ m/s, $\sigma_{n_i} = 9$ dB, $\sigma_{m_i} = 4\frac{\pi}{180}$ rad, $\gamma = 3$, $\gamma_i \sim \mathcal{U}[2.7, 3.3]$, $P_0 = -10$ dBm, $q = 2.5 \times 10^{-3}$ m^2/s^3, $M_c = 1000$.

We present the average RMSE, $\overline{\text{RMSE}}$ (m), performance of the considered algorithms for static sensors setting in both scenarios in Table 3.1. From the table, we can see that the proposed uMAP algorithm performs best in both scenarios, and that the proposed *linearization* technique offers an improvement of roughly 1 m in both scenarios, in comparison with the existing one.

Figure 3.4 illustrates a realization of the estimation process in the first scenario of the proposed (a) uMAP and (b) uKF algorithm, respectively, when sensor mobility is allowed. From hereafter, we only use $N = 2$ mobile sensors, and more particularly the first two sensors from the original setting. From the figure, one can observe that both proposed algorithms solve very efficiently the target tracking problem with only $N = 2$ sensors, owing to their mobility.

Figure 3.5 depicts the RMSE (m) versus t (s) performance comparison of the proposed algorithms in the first scenario for the mobile sensors case. As foreseen, the figure shows the poorest estimation accuracy in the first few time steps, which generally betters with time. This is because, in the first few time instants, the mobile sensors are far away from the target, and as they get closer to it, the performance

Table 3.1 $\overline{\text{RMSE}}$ (m) of the considered algorithms

Algorithm	uMAP	uKF	KF [86]	WLS	MAP	KF
Scenario 1	2.88	3.15	4.31	4.22	2.87	3.13
Scenario 2	2.97	3.22	4.14	4.30	2.97	3.22

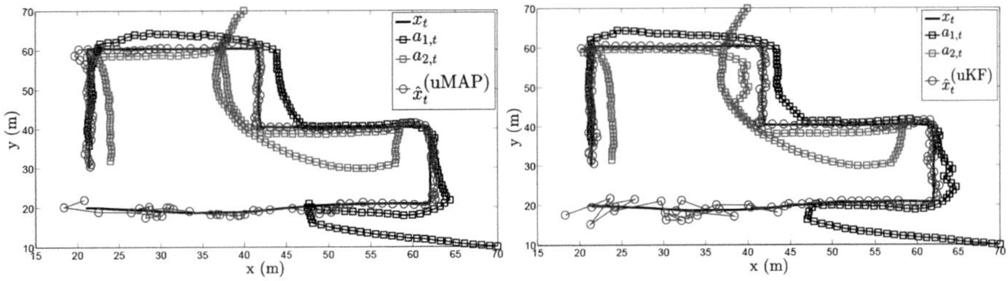

(a) uMAP (b) uKF

Figure 3.4 Illustration of the estimation process in the first scenario, when $N = 2$, $v_a = 1$ m/s, $\sigma_{n_i} = 9$ dB, $\sigma_{m_i} = 4\frac{\pi}{180}$ rad, $\gamma = 3$, $\gamma_i \sim \mathcal{U}[2.7, 3.3]$, $\tau = 5$ m, $P_0 = -10$ dBm, $q = 2.5 \times 10^{-3}$ m^2/s^3.

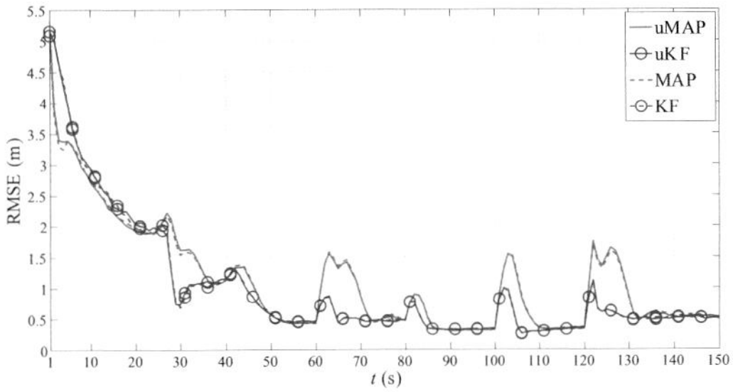

Figure 3.5 RMSE (m) versus t (s) comparison in the first scenario, when $N = 2$, $v_a = 1$ m/s, $\sigma_{n_i} = 9$ dB, $\sigma_{m_i} = 4\frac{\pi}{180}$ rad, $\gamma = 3$, $\gamma_i \sim \mathcal{U}[2.7, 3.3]$, $\tau = 5$ m, $P_0 = -10$ dBm, $q = 2.5 \times 10^{-3}$ m^2/s^3, $M_c = 1000$.

improves in general. Essentially, only at the critical points at which the target takes sharp manoeuvres is where the impairments occur. However, even though we use only $N = 2$ sensors now, due to their mobility, we can see that these deteriorations are notably milder in comparison with the static sensors $N = 3$ case (Figure 3.2). Moreover, the proposed uKF algorithm slightly outperforms the proposed uMAP. Lastly, the new algorithms show exceptional behaviour even for the case where the PLE and the true mobile sensors' locations are not perfectly known.

It might also be of interest for some applications to get an estimate of the target's transmit power. Hence, in Figure 3.6 we show the average ML estimate of P_0, $\widehat{\widehat{P}}_0$ (dBm), in the first scenario through time t (s) for the mobile sensors case. From Figure 3.6, we can see that

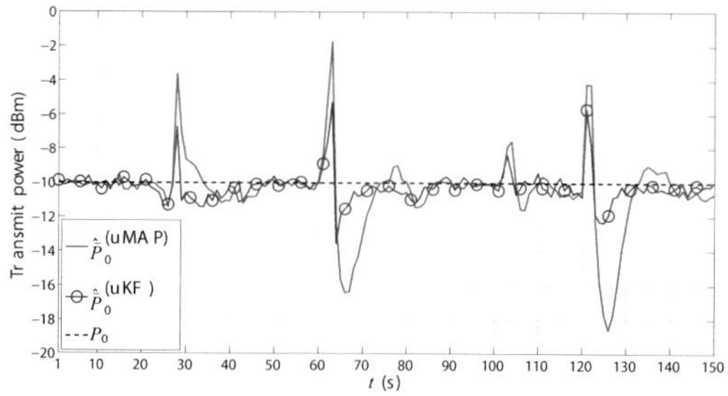

Figure 3.6 $\widehat{\bar{P}}_0$ (dBm) versus t (s) comparison in the first scenario, when $N = 2$, $v_a = 0$ m/s, $\sigma_{n_i} = 9$ dB, $\sigma_{m_i} = 4\frac{\pi}{180}$ rad, $\gamma = 3$, $\gamma_i \sim \mathcal{U}[2.7, 3.3]$, $\tau = 5$ m, $P_0 = -10$ dBm, $q = 2.5 \times 10^{-3}\text{m}^2/\text{s}^3$, $M_c = 1000$.

both proposed algorithms provide an excellent estimate of the transmit power in general. Similar with the case of location estimation, the only significant impairments in the power estimates occur at the critical points.

Figure 3.7 illustrates a realization of the estimation process in the second scenario of the proposed (a) uMAP and (b) uKF algorithm, respectively, when sensor mobility is allowed. As in the first scenario, both proposed algorithms show exceptionally good performance.

Figure 3.8 depicts the RMSE (m) versus t (s) performance comparison of the proposed algorithms in the second scenario for the mobile sensors case. The figure exhibits that both proposed algorithm require a certain amount of time before they *catch up* with the target, after which their estimation performance is outstanding and quite stable.

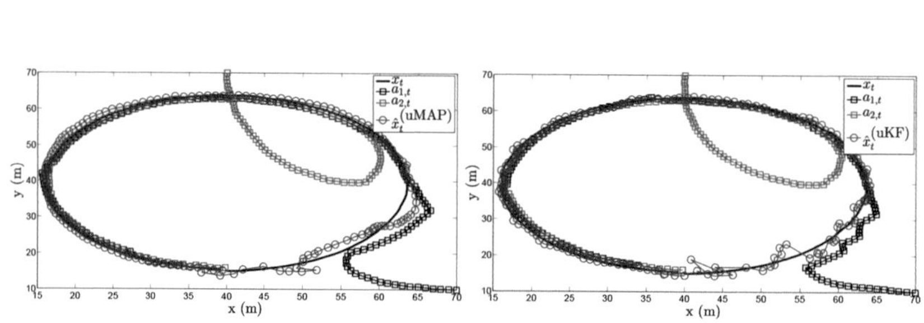

(a) uMAP (b) uKF

Figure 3.7 Illustration of the estimation process in the second scenario, when $N = 2$, $v_a = 1$ m/s, $\sigma_{n_i} = 9$ dB, $\sigma_{m_i} = 4\frac{\pi}{180}$ rad, $\gamma = 3$, $\gamma_i \sim \mathcal{U}[2.7, 3.3]$, $\tau = 5$ m, $P_0 = -10$ dBm, $q = 2.5 \times 10^{-3}\ \text{m}^2/\text{s}^3$.

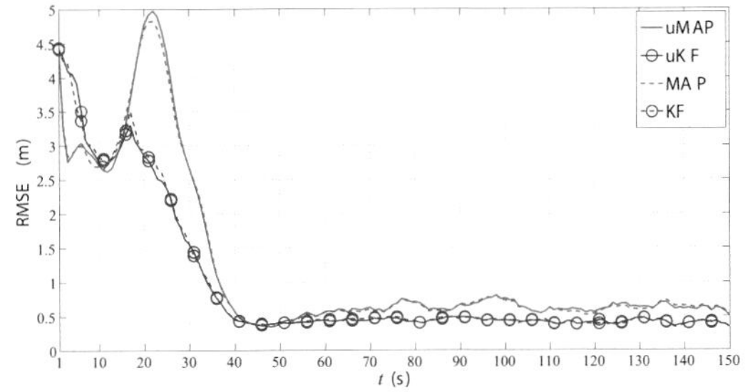

Figure 3.8 RMSE (m) versus t (s) comparison in the second scenario, when $N = 2$, $v_a = 1$ m/s, $\sigma_{n_i} = 9$ dB, $\sigma_{m_i} = 4\frac{\pi}{180}$ rad, $\gamma = 3$, $\gamma_i \sim \mathcal{U}[2.7, 3.3]$, $\tau = 5$ m, $P_0 = -10$ dBm, $q = 2.5 \times 10^{-3}$ m^2/s^3, $M_c = 1000$.

Furthermore, a somewhat better performance of the proposed uKF can be detected in comparison with the uMAP.

In Figure 3.9 we present the $\widehat{\bar{P}}_0$ (dBm) versus t (s) performance comparison in the second scenario for the mobile sensors case. Compared with the results in the first scenario, we can see that the estimation accuracy of P_0 is not as good. This result is interesting on its own, and it seems to be an outcome of the specificity of the target's trajectory (constant change of direction). Nonetheless, the detailed analysis of this phenomenon is beyond the scope of this work. Also, it can be noticed that a considerably better P_0 estimate is obtained through the proposed uKF.

It would also be interesting to investigate the influence of the mobile sensor's velocity on the performance of the proposed algorithms. Consequently, we present the $\overline{\text{RMSE}}$ (m) versus v_a (m/s) performance

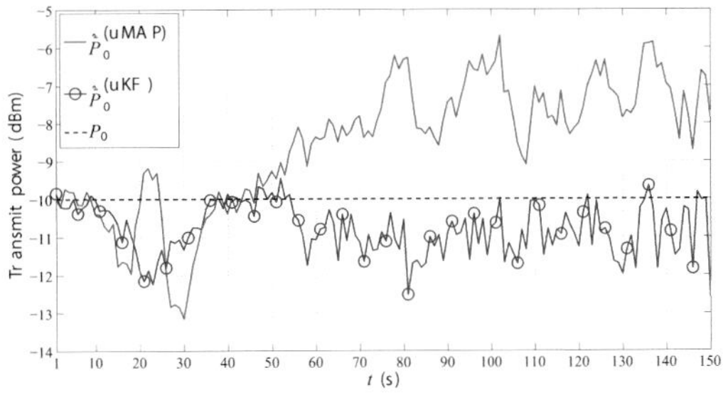

Figure 3.9 $\widehat{\bar{P}}_0$ (dBm) versus t (s) comparison in the second scenario, when $N = 2$, $v_a = 0$ m/s, $\sigma_{n_i} = 9$ dB, $\sigma_{m_i} = 4\frac{\pi}{180}$ rad, $\gamma = 3$, $\gamma_i \sim \mathcal{U}[2.7, 3.3]$, $\tau = 5$ m, $P_0 = -10$ dBm, $q = 2.5 \times 10^{-3}$ m^2/s^3, $M_c = 1000$.

comparison for the first and the second scenario in Figure 3.10 and Figure 3.11, respectively. From the figures, it is obvious that the performance of the proposed algorithms depends on sensors' velocities, and one can notice that the performance of all algorithms betters as v_a (m/s) is increased. This is somewhat intuitive, since the mobile sensors *catch* the target more rapidly as they move at higher velocity. Moreover, the overall performance of the proposed algorithms is very good, while for $v_a \geq \boldsymbol{v}_t$ their performance is remarkable.

3.6.2 Real Indoor Experiment

In this section, we asses the performance of our uMAP tracking algorithm through a real indoor experiment, based on the measurements performed in [67]. Figure 3.12 illustrates the experimental setup of the target tracking scenario. The initial target location is indicated by a red circle and its direction by an arrow. The target passes through the hallways of the building, and at 23 different locations, observations were taken by $N = 7$ anchors, indicated by black squares.

In order to show the advantage of the employed Bayesian approach over the *traditional* ones, Figure 3.13 illustrates the cumulative distribution of the localization error of our uMAP algorithm and the sequential WLS localization algorithm. The figure exhibits clearly the superiority of integrating the prior knowledge into an estimator. Our uMAP offers a median error of mean error (ME) ≈ 2 m, and ME ≤ 3 m in almost 80% of the cases.

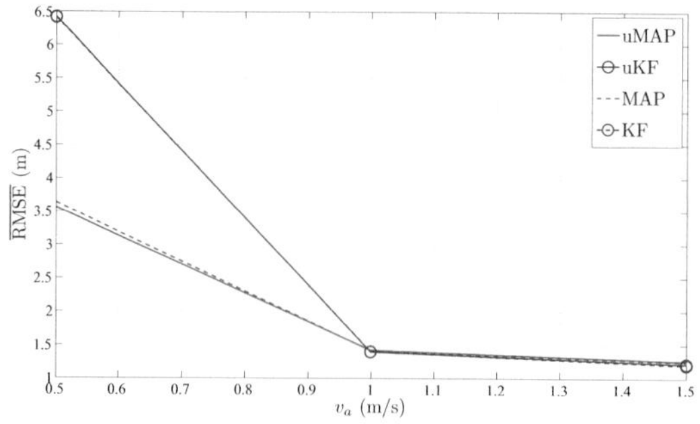

Figure 3.10 $\overline{\text{RMSE}}$ (m) versus v_a (m/s) comparison in the first scenario, when $N = 2$, $\sigma_{n_i} = 9$ dB, $\sigma_{m_i} = 4\frac{\pi}{180}$ rad, $\gamma = 3$, $\gamma_i \sim \mathcal{U}[2.7, 3.3]$, $\tau = 5$ m, $P_0 = -10$ dBm, $q = 2.5 \times 10^{-3}$ m^2/s^3, $M_c = 1000$.

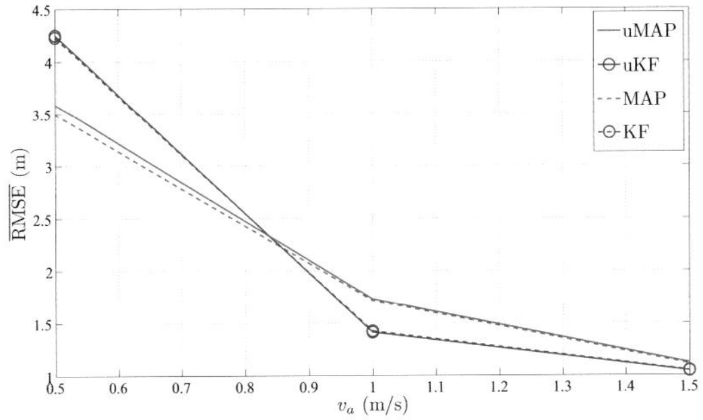

Figure 3.11 $\overline{\mathrm{RMSE}}$ (m) versus v_a (m/s) comparison in the second scenario, when $N = 2$, $\sigma_{n_i} = 9$ dB, $\sigma_{m_i} = 4\frac{\pi}{180}$ rad, $\gamma = 3$, $\gamma_i \sim \mathcal{U}[2.7, 3.3]$, $\tau = 5$ m, $P_0 = -10$ dBm, $q = 2.5 \times 10^{-3}$ m^2/s^3, $M_c = 1000$.

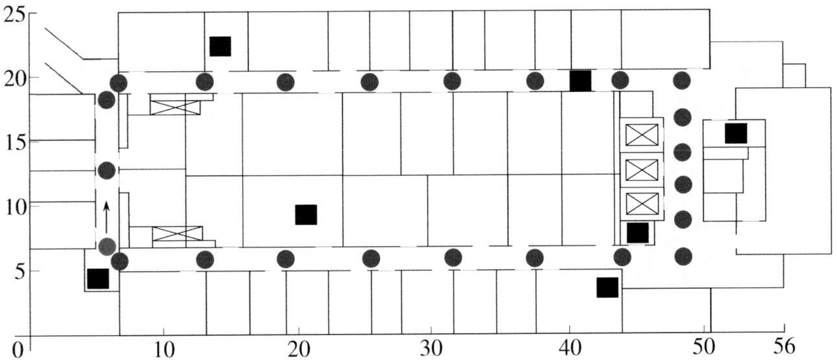

Figure 3.12 Experimental setup for target tracking; the starting point and the direction are indicated by a red circle and an arrow, respectively.

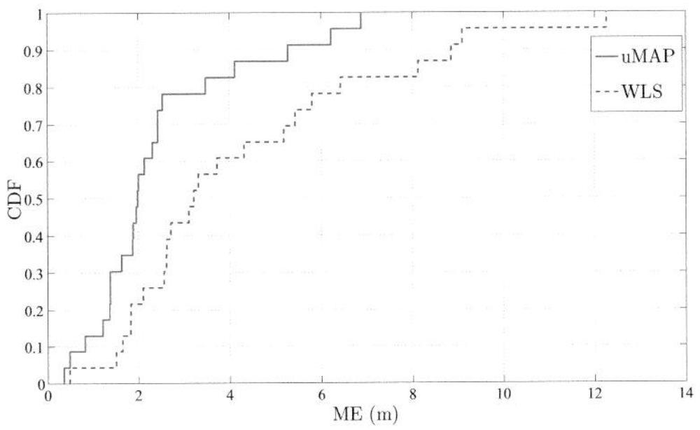

Figure 3.13 Cumulative distribution of the localization error when $N = 7$.

3.7 Conclusions

In this section, we have addressed the target tracking problem in WSN where sensor mobility was granted. The mobile sensors made use of not calibrated RSS measurements with imperfect knowledge about the PLE and unknown target transmit power, which were combined with AoA observations. We have shown that this highly non-linear measurement model can be *linearized* by applying the described procedure. Then, by following the Bayesian methodology, we have managed to integrate the prior knowledge (extracted from the state transition model) with the observations in order to further ameliorate the estimation accuracy. As a result of our work, two novel tracking algorithms were proposed, namely uMAP and uKF. Furthermore, a simple navigation procedure was proposed, which even further improves the estimation accuracy of our algorithms. The new algorithms were compared with the existing KF algorithm and the classical localization algorithm which neglects the prior knowledge in two different scenarios: where the target took sharp manoeuvres and where the target followed a more smooth trajectory. Extensive simulations have been carried out, and the results have confirmed that incorporation of the prior knowledge into an estimator can significantly improve its estimation accuracy. Also, the simulation results showed that the proposed *linearization* technique offers significant error reduction in comparison with the existing one. Moreover, the simulation results corroborated the usefulness of the proposed mobile sensor navigation routine, demonstrating not only a remarkable improvement in the estimation accuracy, but doing so with a reduced number of sensors. Finally, the proposed algorithms exhibited robustness to not knowing transmit power, as well as to imperfect knowledge about the PLE and the true sensors' locations.

4

Conclusions and Future Work

This chapter summarizes the major contributions and attained results of this work (Section 4.1) and discusses, in further detail, some foreseen directions for future research on this topic (Section 4.2).

4.1 Conclusions

This work has studied the problem of target localization in WSN. Two types of networks were considered, non-cooperative and cooperative, for combined RSS and AoA measurements. In the former type of networks, communication between a target and anchors is granted exclusively, while the latter type allows communication between a target and any sensor within its limited communication range. Also, two categories of algorithm conduction were addressed, centralized and distributed. The former category assumes existence of a central node which collects all information gathered in the network and performs all necessary processing, while in the latter category each target determines its own location by making use of local information only. Furthermore, the problem of tracking of a moving target was also investigated. Both cases of target tracking with static anchors and mobile sensors was of interest.

The common objective of the considered localization problems is to estimate the unknown location of the target by solving optimization problems that represent an excellent framework even under inopportune network configuration and strong measurement noise. A strong emphasis was made on convex relaxations and derivation of convex problems, whose global minima can be readily obtained through general-purpose solvers. Furthermore, the solution obtained through the algorithms could also be used as initial point for iterative methods, in which case the risk of convergence to local minima of these methods is minimized, and near-optimal performance can be obtained. The presented algorithms are easy to implement and they offer exceptional estimation accuracy in a single (centralized) or just a few (distributed) iterations in general; thus, making them well suited for practical implementation.

An extensive set of simulation results was presented in the work, together with a detailed analysis of computational complexity. The simulation results corroborate the effectiveness of the presented algorithms, which not only represent an excellent trade-off between estimation accuracy and computational complexity, but also outperform the AoA in terms of estimation accuracy in general. Various network configurations were studied for a broad spectra of parameter settings, and in all of them the presented algorithms showed exceptional performance and robustness to not knowing additional parameters, beyond the target location. Also, it was shown that measurement fusion offers a significant improvement in terms of estimation accuracy in comparison with *traditional* approach. Moreover, it was shown that by exploiting prior knowledge, when it is possible, within an estimator can remarkably boost its performance. Finally, in the case of real-time target tracking, one could have seen that a simple navigation routine, used to manage sensor's movement, can lead to further error reduction, even when a lower number of sensors is employed for that task.

Finally, the performance of the algorithms presented within the work was tested with real indoor experimental data, and the results obtained suggest that our algorithms work well in the considered environment.

4.2 Future Work

There is a number of possibilities for future research. One interesting direction for future research might be development of new and adaptation of the presented algorithms to a more challenging scenarios of indoor localization in severe NLoS environments. NLoS can degrade significantly the localization accuracy, especially in the case where the configuration of the environment is not known, *i.e.*, when it is not known *a priori* which links are LoS and which are NLoS. Instead of trying to distinguish between LoS and NLoS links and disregarding the NLoS ones, because there is always a probability of false alarm or false detection, it would be of interest to exploit the property of positive NLoS bias, which is known to be much larger than the measurement noise.

In this work, a constant network topology during the computational phase was taken for granted. A more realistic scenario, where sensors and/or links can fail with a certain probability might be of interest in some applications, especially for distributed algorithms, which are carried out in an iterative fashion. Such a problem would represent a serious challenge for any localization algorithm, as it could lead to network disconnection or even isolated islands of sensors with no or very scarce information insufficient for *good* location estimation.

Similar to the last possibility, in large-scale WSNs, it might of interest to investigate the case where targets limit the number of cooperating nodes. In the case where a target has a high number of neighbors, we might be interested in selecting only a certain number of its neighbors, by *e.g.*, choosing only the *nearest* ones such that the computational burden is decreased and that its estimation accuracy remains unaffected, or possibly even gets further improved (in the case where one or more very noise links are disregarded). The main challenge in such a problem would be to design an intelligent neighbor-selecting strategy, owing to noisy observations that might mislead a target to disregard a potentially good link and maintain a *bad* one.

In the case of distributed algorithm execution, the design of simple medium access control (MAC) protocols, such as the second-order coloring scheme used in this work, could be interesting as it might lead to error and time-execution reduction. This problem was not in the main scope of the work, and perhaps we did not exploit its full potential. By designing a more intelligent routine for operating hierarchy (*e.g.*, such that targets with the highest number of anchor neighbors work first) might produce better estimation accuracy and at the same time increase the convergence rate of an algorithm, since one would expect to obtain a *better* estimation for those targets which might propagate inside the network.

The work studied the target localization problem by using RSS and combined RSS and AoA measurements. Employing other types of measurements such as ToA, TDoA, frequency or phase of arrival, to name a few, or a combination of them to solve the localization problem might be of interest for future research as well.

For the case of target tracking, this work limited its discussion to tracking of a single target. Tracking of multiple targets simultaneously, possibly with sensor cooperation (if practical interest for such a setting exists), might be another direction for future research. Such a setting represents an extremely challenging problem, since many different aspects of the problem have to be taken into consideration, such as preventing physical collision of sensors, signal interference, and computational complexity of such algorithms to name a few.

Another possible direction for future research might be target navigation. By knowing the terrain configuration and by tracking the location of a mobile target, a relatively accurate target navigation could be done. Such an application might be of practical interest in search and rescue missions, exploration in hostile environments and robotics.

This work assumed omnidirectional antenna directivity such that the set of all possible solutions belongs to the area formed by an intersection of multiple circle-shaped contours. Although the presented methods work well in all considered scenarios, this assumption might be an oversimplification of the problem, since the antenna radiation pattern is non-isotropic in practice (*e.g.*, antenna radiation pattern depends on antenna geometry configuration – shape and dimension, dielectric material, combination (antenna array), and signal wavelength). Therefore, in practice, the set of all possible solutions belongs to the area formed by intersection of non-circular power contours determined by the antenna pattern. Hence, there seems to be some room for further improvement of the presented algorithms by taking the antenna pattern into consideration when deriving a localization scheme.

Lastly, validation of all potential algorithms and ideas for the described outline through experimental setup would be of great interest. In this work, we have used the measurements and scenario from another work, owing to unselfishness and kindness of our colleagues from Rutgers University. However, by doing so, we were very limited to that specific scenario and many times we were in doubt regarding the measurement setting. Thus, it would be of great personal interest for us to conduct an experiment by ourselves and be completely autonomous when validating our ideas.

A

CRB Derivation for RSS-AoA Localization

CRB provides a lower bound on the variance of any unbiased estimator, meaning that it is physically impossible to find an unbiased estimator whose variance is less than the bound. CRB offers us a benchmark against which we can compare the performance of any unbiased estimator. If the estimator attains the bound for all values of the unknown parameters, we say that such estimator is the minimum variance unbiased estimator [46].

Let $\boldsymbol{y} = [\boldsymbol{x}_k^T, P_0]^T, k = 1, ..., M$, denote the $3M + 1$ vector of all unknown parameters. According to [46], the variance of any unbiased estimator is lower bounded by $\text{var}(\hat{\boldsymbol{y}}) \geq [\boldsymbol{J}^{-1}(\boldsymbol{y})]_{ii}$, where $\boldsymbol{J}(\boldsymbol{y})$ is the $(3M+1) \times (3M+1)$ Fisher information matrix (FIM). The elements of the FIM are defined as $[\boldsymbol{J}(\boldsymbol{y})]_{i,j} = -\mathbb{E}\left[\frac{\partial^2 \ln p(\boldsymbol{\theta}|\boldsymbol{y})}{\partial \boldsymbol{y}_i \partial \boldsymbol{y}_j}\right]$, where $i, j = 1, ..., (3M+1)$, and $p(\boldsymbol{\theta}|\boldsymbol{y})$ is the joint conditional probability density function of the observation vector $\boldsymbol{\theta} = [\boldsymbol{P}^T, \boldsymbol{\phi}^T, \boldsymbol{\alpha}^T]$ ($\boldsymbol{P} = [P_{ij}^{\mathcal{A}}, P_{ik}^{\mathcal{B}}]^T$, $\boldsymbol{\phi} = [\phi_{ij}^{\mathcal{A}}, \phi_{ik}^{\mathcal{B}}]^T$, $\boldsymbol{\alpha} = [\alpha_{ij}^{\mathcal{A}}, \alpha_{ik}^{\mathcal{B}}]^T$), given $\boldsymbol{y}$.

Then, the FIM is computed as:

$$\begin{aligned} \boldsymbol{J}(\boldsymbol{y}) = & \frac{1}{\sigma_{n_{ij}}^2} \sum_{(i,j):(i,j)\in\mathcal{A}} \boldsymbol{h}_{ij}\boldsymbol{h}_{ij}^T + \frac{1}{\sigma_{m_{ij}}^2} \sum_{(i,j):(i,j)\in\mathcal{A}} \boldsymbol{q}_{ij}\boldsymbol{q}_{ij}^T \\ & + \frac{1}{\sigma_{v_{ij}}^2} \sum_{(i,j):(i,j)\in\mathcal{A}} \boldsymbol{u}_{ij}\boldsymbol{u}_{ij}^T + \frac{1}{\sigma_{n_{ik}}^2} \sum_{(i,k):(i,k)\in\mathcal{B}} \boldsymbol{h}_{ik}\boldsymbol{h}_{ik}^T \\ & + \frac{1}{\sigma_{m_{ik}}^2} \sum_{(i,k):(i,k)\in\mathcal{B}} \boldsymbol{q}_{ik}\boldsymbol{q}_{ik}^T + \frac{1}{\sigma_{v_{ik}}^2} \sum_{(i,k):(i,k)\in\mathcal{B}} \boldsymbol{u}_{ik}\boldsymbol{u}_{ik}^T, \end{aligned}$$

where

$$\boldsymbol{h}_{ij} = \boldsymbol{\rho} - \frac{10\gamma d_0}{\ln(10)} \frac{\boldsymbol{E}_i(\boldsymbol{E}_i^T\boldsymbol{y} - \boldsymbol{a}_j)}{\|\boldsymbol{E}_i^T\boldsymbol{y} - \boldsymbol{a}_j\|^2},$$

$$\boldsymbol{q}_{ij} = \frac{\boldsymbol{E}_i \boldsymbol{e}_2(\boldsymbol{e}_1^T \boldsymbol{E}_i^T \boldsymbol{y} - \boldsymbol{e}_1^T \boldsymbol{a}_j) - \boldsymbol{E}_i \boldsymbol{e}_1(\boldsymbol{e}_2^T \boldsymbol{E}_i^T \boldsymbol{y} - \boldsymbol{e}_2^T \boldsymbol{a}_j)}{(\boldsymbol{e}_1^T \boldsymbol{E}_i^T \boldsymbol{y} - \boldsymbol{e}_1^T \boldsymbol{a}_j)^2 + (\boldsymbol{e}_2^T \boldsymbol{E}_i^T \boldsymbol{y} - \boldsymbol{e}_2^T \boldsymbol{a}_j)^2},$$

$$\boldsymbol{u}_{ij} = \frac{\boldsymbol{E}_i(\boldsymbol{e}_3 \|\boldsymbol{E}_i^T \boldsymbol{y} - \boldsymbol{a}_j\| - (\boldsymbol{E}_i^T \boldsymbol{y} - \boldsymbol{a}_j)(\boldsymbol{e}_3^T \boldsymbol{E}_i^T \boldsymbol{y} - \boldsymbol{e}_3^T \boldsymbol{a}_j))}{\|\boldsymbol{E}_i^T \boldsymbol{y} - \boldsymbol{a}_j\|^2 \sqrt{\|\boldsymbol{E}_i^T \boldsymbol{y} - \boldsymbol{a}_j\|^2 - (\boldsymbol{e}_3^T \boldsymbol{E}_i^T \boldsymbol{y} - \boldsymbol{e}_3^T \boldsymbol{a}_j)^2}},$$

$$\boldsymbol{h}_{ik} = \boldsymbol{\rho} - \frac{10\gamma d_0}{\ln(10)} \frac{(\boldsymbol{E}_i - \boldsymbol{E}_k)(\boldsymbol{E}_i^T \boldsymbol{y} - \boldsymbol{E}_k^T \boldsymbol{y})}{\|\boldsymbol{E}_i^T \boldsymbol{y} - \boldsymbol{E}_k^T \boldsymbol{y}\|^2},$$

$$\boldsymbol{q}_{ik} = \frac{(\boldsymbol{E}_i - \boldsymbol{E}_k)(\boldsymbol{e}_2(\boldsymbol{e}_1^T \boldsymbol{E}_i^T \boldsymbol{y} - \boldsymbol{e}_1^T \boldsymbol{E}_k^T \boldsymbol{y}) - \boldsymbol{e}_1(\boldsymbol{e}_2^T \boldsymbol{E}_i^T \boldsymbol{y} - \boldsymbol{e}_2^T \boldsymbol{E}_k^T \boldsymbol{y}))}{(\boldsymbol{e}_1^T \boldsymbol{E}_i^T \boldsymbol{y} - \boldsymbol{e}_1^T \boldsymbol{E}_k^T \boldsymbol{y})^2 + (\boldsymbol{e}_2^T \boldsymbol{E}_i^T \boldsymbol{y} - \boldsymbol{e}_2^T \boldsymbol{E}_k^T \boldsymbol{y})^2},$$

$$\boldsymbol{u}_{ik} = \frac{(\boldsymbol{E}_i - \boldsymbol{E}_k)(\boldsymbol{e}_3 \|\boldsymbol{E}_i^T \boldsymbol{y} - \boldsymbol{E}_k^T \boldsymbol{y}\| - (\boldsymbol{E}_i^T \boldsymbol{y} - \boldsymbol{E}_k^T \boldsymbol{y})(\boldsymbol{e}_3^T \boldsymbol{E}_i^T \boldsymbol{y} - \boldsymbol{e}_3^T \boldsymbol{E}_k^T \boldsymbol{y}))}{\|\boldsymbol{E}_i^T \boldsymbol{y} - \boldsymbol{E}_k^T \boldsymbol{y}\|^2 \sqrt{\|\boldsymbol{E}_i^T \boldsymbol{y} - \boldsymbol{E}_k^T \boldsymbol{y}\|^2 - (\boldsymbol{e}_3^T \boldsymbol{E}_i^T \boldsymbol{y} - \boldsymbol{e}_3^T \boldsymbol{E}_k^T \boldsymbol{y})^2}},$$

and $\boldsymbol{\rho} = [\boldsymbol{0}_{1\times 3M}\, , 1]^T$, $\boldsymbol{E}_i = [\boldsymbol{e}_{3i-2},\, \boldsymbol{e}_{3i-1},\, \boldsymbol{e}_{3i}]$, where $\boldsymbol{e}_i$ represents the i-th column of the identity matrix $\boldsymbol{I}_{3M+1}$, and $\boldsymbol{e}_1 = [1,\, 0,\, 0]^T$, $\boldsymbol{e}_2 = [0,\, 1,\, 0]^T$ and $\boldsymbol{e}_3 = [0,\, 0,\, 1]^T$.

Therefore, the CRB for the estimate of the target positions is computed as:

$$\text{CRB} = \text{trace}\left(\left[\boldsymbol{J}^{-1}(\boldsymbol{y})\right]_{1:3M,1:3M}\right),$$

where $[\boldsymbol{M}]_{a:b,c:d}$ represents the sub-matrix of the matrix $\boldsymbol{M}$ composed of the rows a to b and the columns c to d of M.

B

Derivation of the State Transition Model

Consider the following continuous-time state transition model [94].

$$\dot{\boldsymbol{\theta}}(t) = \boldsymbol{A}\boldsymbol{\theta}(t) + \boldsymbol{D}\boldsymbol{u}(t) + \boldsymbol{B}\boldsymbol{r}(t), \quad \boldsymbol{\theta}(t_0) = \boldsymbol{\theta}_0, \tag{B.1}$$

where $\boldsymbol{\theta}(t) \in \mathbb{R}^n$ is the state vector, $\boldsymbol{u}(t) \in \mathbb{R}^p$ is the vector containing any control inputs (steering angle, throttle setting, breaking force), $\boldsymbol{A} \in \mathbb{R}^{n\times n}$, $\boldsymbol{D} \in \mathbb{R}^{n\times p}$ and $\boldsymbol{B} \in \mathbb{R}^{n\times r}$ are the transition, input gain and noise gain matrices, respectively, and $\boldsymbol{r}(t)$ is a continuous-time process noise with covariance $\boldsymbol{Q}(t)$.

By using Euler's method or zero-order hold [95, 96], we can rewrite the continuous-time state transition model (B.1) for a time-invariant continuous-time system with sampling rate Δ, for initial time $t_0 = t\Delta$ and final time $t_f = (t+1)\Delta$, as

$$\boldsymbol{\theta}(t_f) = \exp\{\boldsymbol{A}(t_f - t_0)\}\boldsymbol{\theta}(t_0) + \int_{t_0}^{t_f} \exp\{\boldsymbol{A}(t_f - \tau)\}\left(\boldsymbol{D}\boldsymbol{u}(\tau) + \boldsymbol{B}\boldsymbol{r}(\tau)\right) d\tau,$$

which is equivalent to

$$\boldsymbol{\theta}(t+1) = \exp\{\boldsymbol{A}\Delta\}\boldsymbol{\theta}(t) + \int_{t\Delta}^{(t+1)\Delta} \exp\{\boldsymbol{A}((t+1)\Delta - \tau)\}\left(\boldsymbol{D}\boldsymbol{u}(\tau) + \boldsymbol{B}\boldsymbol{r}(\tau)\right) d\tau. \tag{B.2}$$

If we assume that the input $\boldsymbol{u}(t)$ changes slowly, relatively to the sampling period, we have $\boldsymbol{u}(t_f) \approx \boldsymbol{u}(t_0)$ for $t_0 \leq t \leq t_f$. Then by changing the variable of integration $\varphi = (t+1)\Delta - \tau$ such that $d\varphi = -d\tau$, (B.2) can be rewritten

$$\begin{aligned}\boldsymbol{\theta}(t+1) = \exp\{\boldsymbol{A}\Delta\}\boldsymbol{\theta}(t) &+ \int_{\Delta}^{0} \exp\{\boldsymbol{A}\varphi\}\boldsymbol{D}(-d\varphi)\boldsymbol{u}(t) \\ &+ \int_{t\Delta}^{(t+1)\Delta} \exp\{\boldsymbol{A}((t+1)\Delta - \tau)\}\boldsymbol{B}\boldsymbol{r}(\tau)d\tau\end{aligned}$$

$$= \exp\{\boldsymbol{A}\Delta\}\boldsymbol{\theta}(t) + \int_0^{\Delta} \exp\{\boldsymbol{A}\varphi\}\boldsymbol{D}d\varphi\boldsymbol{u}(t)$$
$$+ \int_{t\Delta}^{(t+1)\Delta} \exp\{\boldsymbol{A}((t+1)\Delta - \tau)\}\boldsymbol{B}\boldsymbol{r}(\tau)d\tau, \tag{B.3}$$

and the state model (B.1) can be *discretized* as

$$\boldsymbol{\theta}_{t+1} = \boldsymbol{S}\boldsymbol{\theta}_t + \boldsymbol{G}\boldsymbol{u}_t + \boldsymbol{r}_t,$$

where

$$\boldsymbol{S} = \exp\{\boldsymbol{A}\Delta\}, \tag{B.4}$$
$$\boldsymbol{G} = \int_0^{\Delta} \exp\{\boldsymbol{A}\varphi\}\boldsymbol{D}d\varphi,$$
$$\boldsymbol{r}_t = \int_{t\Delta}^{(t+1)\Delta} \exp\{\boldsymbol{A}((t+1)\Delta - \tau)\}\boldsymbol{B}\boldsymbol{r}(\tau)d\tau. \tag{B.5}$$

The process noise, $\boldsymbol{r}(t)$, is assumed to be zero-mean and white Gaussian, and the *discretized* process noise, $\boldsymbol{r}_t$, retains the same characteristics [94], *i.e.*,

$$E[\boldsymbol{r}_t] = 0, \quad E\left[\boldsymbol{r}_t\boldsymbol{r}_t^T\right] = \boldsymbol{Q}_t\delta_t,$$

where δ_t represents a Dirac impulse, and the covariance of the state process noise is given, according to (B.5), as

$$\boldsymbol{Q}_t = \int_0^{\Delta} \exp\{\boldsymbol{A}((t+1)\Delta - \tau)\}\boldsymbol{B}\boldsymbol{Q}\boldsymbol{B}^T \exp\left\{\boldsymbol{A}^T((t+1)\Delta - \tau)\right\}d\tau, \tag{B.6}$$

and $\boldsymbol{Q} = \mathrm{diag}([q, q])$, with q denoting a tuning parameter for the state process noise intensity.

Since this work assumes a 2-dimensional constant velocity model, the continuous-time target state model (B.1) can be simplified [94] as

$$\dot{\boldsymbol{\theta}}(t) = \boldsymbol{A}\boldsymbol{\theta}(t) + \boldsymbol{B}\boldsymbol{r}(t), \tag{B.7}$$

where

$$\boldsymbol{A} = \begin{bmatrix} 0 & 0 & 1 & 0 \\ 0 & 0 & 0 & 1 \\ 0 & 0 & 0 & 0 \\ 0 & 0 & 0 & 0 \end{bmatrix}, \quad \boldsymbol{B} = \begin{bmatrix} 0 & 0 \\ 0 & 0 \\ 1 & 0 \\ 0 & 1 \end{bmatrix},$$

and $\boldsymbol{r}(t) \sim \mathcal{N}(\boldsymbol{0}, \boldsymbol{Q})$.

The discrete-time model equivalent to the above one is described by

$$\boldsymbol{\theta}_{t+1} = \boldsymbol{S}\boldsymbol{\theta}_t + \boldsymbol{r}_t, \tag{B.8}$$

where, by solving (B.4) and (B.6) respectively, we get

$$\boldsymbol{S} = \begin{bmatrix} 1 & 0 & \Delta & 0 \\ 0 & 1 & 0 & \Delta \\ 0 & 0 & 1 & 0 \\ 0 & 0 & 0 & 1 \end{bmatrix}, \quad \boldsymbol{Q} = q \begin{bmatrix} \frac{\Delta^3}{3} & 0 & \frac{\Delta^2}{2} & 0 \\ 0 & \frac{\Delta^3}{3} & 0 & \frac{\Delta^2}{2} \\ \frac{\Delta^2}{2} & 0 & \Delta & 0 \\ 0 & \frac{\Delta^2}{2} & 0 & \Delta \end{bmatrix}.$$

Bibliography

[1] N. Patwari, "Location estimation in sensor networks," PhD thesis, University of Michigan, Jul. 2005.
[2] Y. Singh, S. Saha, U. Chugh, and C. Gupt, "Distributed event detection in wireless sensor networks for forest fires," in *UKSim*, Cambridge, UK, Apr. 2013, pp. 634–639.
[3] Z. Rongbai and C. Guohua, "Research on major hazard installations monitoring system based on wsn," in *ICFCC*, Wuhan, Hubei, China, May 2010, pp. V1-741–V1-745.
[4] Z. Dai, S. Wang, and Z. Yan, "Bshm-wsn: A wireless sensor network for bridge structure health monitoring," in *ICMIC*, Wuhan, Hubei, China, Jun. 2012, pp. 708–712.
[5] L. Blazevic, J. Y. Le Boudec, and S. Giordano, "A location-based routing method for mobile ad hoc networks," *IEEE Trans. Mobile Computing*, vol. 4, no. 2, pp. 97–110, 2005.
[6] L. Ghelardoni, A. Ghio, and D. Anguita, "Smart underwater wireless sensor networks," in *IEEEI*, Eilat, Israel, Nov. 2012, pp. 1–5.
[7] T. He, S. Krishnamurthy, J. A. Stankovic, T. Abdelzaher, L. Luo, R. Stoleru, T. Yan, and L. Gu, "Energy-efficient surveillance system using wireless sensor networks," in *MobiSys*, Boston, MA, USA, Jun. 2004, pp. 1–14.
[8] M. D. Dikaiakos, A. Florides, T. Nadeem, and L. Iftode, "Location-aware services over vehicular ad-hoc networks using car-to-car communication," *IEEE J. Selected Areas in Commun.*, vol. 25, no. 8, pp. 1590–1602, 2007.
[9] M. Faschinger, C. R. Sastry, A. H. Patel, and N. C. Tas, "An rfid and wireless sensor network-based implementation of workflow optimization," in *WoWMoM*, Helsinki, Finland, Jun. 2007, pp. 1–8.

[10] N. Deshpande, E. Grant, and T. C. Henderson, "Target-directed navigation using wireless sensor networks and implicit surface interpolation," in *ICRA*, Saint Paul, Minnesota, USA, May 2012, pp. 457–462.

[11] E. Xu, Z. Ding and S. Dasgupta, "Target tracking and mobile sensor navigation in wireless sensor networks," *IEEE Trans. Mobile Comput.*, vol. 12, no. 1, pp. 177–186, 2013.

[12] N. Deshpande, E. Grant, and T. C. Henderson, "Target localization and autonomous navigation using wireless sensor networks–a pseudogradient algorithm approach," *IEEE Systems Journal*, vol. 8, no. 1, pp. 93–103, 2014.

[13] P. G. Kandhare and G. M. Bhandari, "Guidance providing navigation in target tracking for wireless sensor networks," *IJSR*, vol. 4, no. 6, pp. 2795–2798, 2015.

[14] Z. Sahinoglu, S. Gezici, and I. Güvenc, *Ultra-wideband positioning systems: Theoretical limits, ranging algorithms, and protocols*, 1st Ed. Cambridge University Press, NY, USA, 2008.

[15] L. Buttyán and J. P. Hubaux, *Security and cooperation in wireless networks: Thwarting malicious and selfish behavior in the age of ubiquitous computing*, 1st Ed. Cambridge University Press, NY, USA, 2007.

[16] N. Patwari, J. N. Ash, S. Kyperountas, A. O Hero III, R. L. Moses, and N. S. Correal, "Locating the nodes: Cooperative localization in wireless sensor networks," *IEEE Signal Process. Mag.*, vol. 22, no. 4, pp. 54–69, 2005.

[17] F. Bandiera, A. Coluccia, and G. Ricci, "A cognitive algorithm for received signal strength based localization," *IEEE Trans. Signal Process.*, vol. 63, no. 7, pp. 1726–1736, 2015.

[18] G. Destino, "Positioning in wireless networks: Noncooperative and cooperative algorithms," PhD thesis, University of Oulu, Oct. 2012.

[19] J. He, Y. Geng, and K. Pahlavan, "Toward accurate human tracking: Modeling time-of-arrival for wireless wearable sensors in multipath environment," *IEEE Sensors Journal*, vol. 14, no. 11, pp. 3996–4006, 2014.

[20] X. Qua and L. Xie, "An efficient convex constrained weighted least squares source localization algorithm based on tdoa measurements," *Elsevier Sign. Process.*, vol. 16, no. 119, pp. 142–152, 2016.

[21] J. Cota-Ruiz, J. G. Rosiles, P. Rivas-Perea, and E. Sifuentes, "A distributed localization algorithm for wireless sensor networks

based on the solution of spatially-constrained local problems," *IEEE Sensors Journal*, vol. 13, no. 6, pp. 2181–2191, 2013.

[22] Y. Wang and K. C. Ho, "An asymptotically efficient estimator in closed-form for 3d aoa localization using a sensor network," *IEEE Trans. Wirel. Commun.*, vol. 14, no. 12, pp. 6524–6535, 2015.

[23] S. Tomic, M. Beko, and R. Dinis, "Distributed rss-based localization in wireless sensor networks based on second-order cone programming," *MDPI Sensors*, vol. 14, no. 10, pp. 18 410–18 432, 2014.

[24] N. Salman, M. Ghogho, and A. H. Kemp, "Optimized low complexity sensor node positioning in wireless sensor networks," *IEEE Sensors Journal*, vol. 14, no. 1, pp. 39–46, 2014.

[25] N. Bulusu, J. Heidemann, and D. Estrin, "Gps-less low cost outdoor localization for very small devices," *IEEE Personal Commun. Mag.*, vol. 7, no. 5, pp. 28–34, 2000.

[26] A. Bahillo, S. Mazuelas, R. M. Lorenzo, P. Fernández, J. Prieto, R. J. Durán, and E. J. Abril, "Hybrid rss-rtt localization scheme for indoor wireless networks," *EURASIP J. Advances in Sign. Process.*, vol. 10, no. 1, pp. 1–12, 2010.

[27] N. Alam and A. G. Dempster, "Cooperative positioning for vehicular networks: Facts and future," *IEEE Trans. Intelligent Transportation Systems*, vol. 14, no. 9, pp. 1708–1717, 2013.

[28] U. Hatthasin, S. Thainimit, K. Vibhatavanij, N. Premasathian, and D. Worasawate, "The use of rtof and rss for a one base station rfid system," *IJCSNS*, vol. 10, no. 7, pp. 184–195, 2010.

[29] T. Gädeke, J. Schmid, J. J. M. Krüge, W. Stork, and K. D. Müller-Glaser, "A bi-modal ad-hoc localization scheme for wireless networks based on rss and tof fusion," in *WPNC*, Dresden, Germany, Mar. 2013, pp. 1–6.

[30] K. Yu, "3-d localization error analysis in wireless networks," *IEEE Trans. Wireless Commun.*, vol. 6, no. 10, pp. 3473–3481, 2007.

[31] S. Wang, B. R. Jackson, and R. Inkol, "Hybrid rss/aoa emitter location estimation based on least squares and maximum likelihood criteria," in *IEEE QBSC*, Kingston, ON, Canada, Jun. 2012, pp. 24–29.

[32] L. Gazzah, L. Najjar, and H. Besbes, "Selective hybrid rss/aoa weighting algorithm for nlos intra cell localization," in *IEEE WCNC*, Istanbul, Turkey, Apr. 2014, pp. 2546–2551.

[33] Y. T. Chan, F. Chan, W. Read, B. R. Jackson, and B. H. Lee, "Hybrid localization of an emitter by combining angle-of-arrival and received signal strength measurements," in *IEEE CCECE*, Toronto, ON, Canada, May 2014, pp. 1–5.

[34] S. Tomic, M. Marikj, M. Beko, R. Dinis, and N. Orfao, "Hybrid rss-aoa technique for 3-d node localization in wireless sensor networks," in *IEEE IWCMC*, Dubrovnik, Croatia, Apr. 2015, pp. 1277–1282.

[35] C. Cheng, W. Hu, and W. P. Tay, "Localization of a moving non-cooperative rf target in nlos environment using rss and aoa measurements," in *IEEE ICASSP*, South Brisbane, Queensland, Australia, Apr. 2015, pp. 3581–3585.

[36] S. Tomic, M. Beko, and R. Dinis, "3-d target localization in wireless sensor network using rss and aoa measurement," *IEEE Trans. Vehic. Technol.*, vol. 66, no. 4, pp. 3197–3210, 2017.

[37] A. Singh, S. Kumar, and O. Kaiwartya, "A hybrid localization algorithm for wireless sensor networks," in *ICRTC*, Ghaziabad India, Mar. 2015, pp. 1432–1439.

[38] S. Tomic, M. Beko, and R. Dinis, "Distributed rss-aoa based localization with unknown transmit powers," *IEEE Wirel. Commun. Letters*, vol. 5, no. 4, pp. 392–395, 2016.

[39] S. Sundhar Ram, A. Nedić, and V. V. Veeravalli, "Distributed subgradient projection algorithm for convex optimization," in *IEEE ICASSP*, Taipei, Taiwan, Apr. 2009, pp. 3653–3656.

[40] I. Guvenc and C. C. Chong, "A survey on toa based wireless localization and nlos mitigation techniques," *IEEE Commun. Survey and Tutorials*, vol. 11, no. 3, pp. 107–124, 2009.

[41] J. Blumenthal, R. Grossmann, F. Golatowski, and D. Timmermann, "Weighted centroid localization in zigbee-based sensor networks," in *WISP*, Alcala de Henares, Spain, Oct. 2007, pp. 1–6.

[42] M. Brunato and R. Battiti, "Statistical learning theory for location fingerprinting in wireless LANs," *Elsevier, J. Computer Networks*, vol. 47, no. 6, pp. 825–845, 2005.

[43] D. E. Manolakis, "Efficient solution and performance analysis of 3-d position estimation by trilateration," *IEEE Trans. Aerospace and Electronic Systems*, vol. 32, no. 4, pp. 1239–1248, 1996.

[44] A. H. Sayed, A. Tarighat, and N. Khajehnouri, "Network-based wireless location," *IEEE Sign. Process. Mag.*, vol. 22, no. 40, pp. 24–40, 2005.

[45] B. T. Fang, "Simple solutions for hyperbolic and related position fixes," *IEEE Trans. Aerospace and Electronic Systems*, vol. 26, no. 5, pp. 748–758, 1990.

[46] S. M. Kay, *Fundamentals of Statistical Signal Processing: Estimation theory*, 1st Ed. Prentice Hall Upper Saddle River, NJ, USA, 1993.

[47] X. Li, "Collaborative localization with received-signal strength in wireless sensor networks," *IEEE Trans. Veh. Technol.*, vol. 56, no. 6, pp. 3807–3817, 2007.

[48] K. W. K. Lui, W. K. Ma, H. C. So, and F. K. W. Chan, "Semi-definite programming algorithms for sensor network node localization with uncertainties in anchor positions and/or propagation speed," *IEEE Trans. Signal Process.*, vol. 57, no. 2, pp. 752–763, 2009.

[49] R. W. Ouyang, A. K. S. Wong, and C. T. Lea, "Received signal strength-based wireless localization via semidefinite programming: Noncooperative and cooperative schemes," *IEEE Trans. Veh. Technol.*, vol. 59, no. 3, pp. 1307–1318, 2010.

[50] G. Wang and K. Yang, "A new approach to sensor node localization using rss measurements in wireless sensor networks," *IEEE Trans. Wireless Commun.*, vol. 10, no. 5, pp. 1389–1395, 2011.

[51] R. M. Vaghefi, M. R. Gholami, and E. G. Ström, "Rss-based sensor localization with unknown transmit power," in *IEEE ICASSP*, Prague, Czech Republic, Apr. 2011, pp. 2480–2483.

[52] H. Chen, G. Wang, Z. Wang, H. C. So, and H. V. Poor, "Non-line-of-sight node localization based on semi-definite programming in wireless sensor networks," *IEEE Trans. Wireless Commun.*, vol. 11, no. 1, pp. 108–116, 2012.

[53] G. Wang, H. Chen, Y. Li, and M. Jin, "On received-signal-strength based localization with unknown transmit power and path loss exponent," *IEEE Wireless Commun. Letters*, vol. 1, no. 5, pp. 536–539, 2012.

[54] S. Tomic, M. Beko, R. Dinis, and V. Lipovac, "Rss-based localization in wireless sensor networks using socp relaxation," in *IEEE ICASSP*, Darmstadt, Germany, Jun. 2013, pp. 749–753.

[55] R. M. Vaghefi, R. M. B. M. R. Gholami, and E. G. Ström, "Cooperative received signal strength-based sensor localization with unknown

transmit powers," *IEEE Trans. Signal Process.*, vol. 61, no. 6, pp. 1389–1403, 2013.

[56] S. Boyd and L. Vandenberghe, *Convex optimization*, 1st Ed. Cambridge University Press, Cambridge, UK, 2004.

[57] S. Tomic, M. Beko, and R. Dinis, "Rss-based localization in wireless sensor networks using convex relaxation: Noncooperative and cooperative schemes," *IEEE Trans. Veh. Technol.*, vol. 64, no. 5, pp. 2037–2050, 2015.

[58] P. Biswas, T. C. Lian, T. C. Wang, and Y. Ye, "Semidefinite programming based algorithms for sensor network localization," *ACM Trans. Sensor Networks*, vol. 2, no. 2, pp. 188–220, 2006.

[59] R. Fletcher, *Practical methods of optimization*, 1st Ed. John Wiley & Sons, Chichester, UK, 1987.

[60] P. Biswas, H. Aghajan, and Y. Ye, "Semidefinite programming algorithms for sensor network localization using angle of arrival information," in *Asilomar Conference on Signals, Systems, and Computers*, Pacific Grove, CA, USA, Oct. 2005, pp. 220–224.

[61] T. S. Rappaport, *Wireless communications: Principles and practice*, 1st Ed. Prentice Hall Upper Saddle River NJ, USA, 1996.

[62] M. L. Sichitiu and V. Ramadurai, "Localization of wireless sensor networks with a mobile beacon," in *IEEE MASS*, Fort Lauderdale, FL, USA, Oct. 2004, pp. 174–183.

[63] M. B. Ferreira, "*Hybrid indoor localization based on ranges and video*," MSc thesis, Universidade Técnica de Lisboa, 2014.

[64] M. B. Ferreira, J. Gomes, and J. P. Costeira, "A unified approach for hybrid source localization based on ranges and video," in *IEEE ICASSP*, South Brisbane, Queensland, Australia, Apr. 2015, pp. 2879–2883.

[65] M. B. Ferreira, J. Gomes, C. Soares, and J. P. Costeira, "Collaborative localization of vehicle formations based on ranges and bearings," in *UComms*, Lerici, Italy, Apr. 2016, pp. 1–5.

[66] Z. Xiang and U. Ozguner, "A 3-d positioning system for off-road autonomous vehicles," in *IEEE IV*, Las Vegas, NV, USA, Jun. 2005, pp. 130–134.

[67] D. Niculescu and B. Nath, "Vor base stations for indoor 802.11 positioning," in *ACM MobiCom*, Philadelphia, PA, USA, Sep. 2004, pp. 58–69.

[68] I. J. Myung, "Tutorial on maximum likelihood estimation," *J. Mathematical Psychology*, vol. 47, no. 1, pp. 90–100, 2003.

[69] A. Beck, P. Stoica, and J. Li, "Exact and approximate solutions of source localization problems," *IEEE Trans. Signal Process.*, vol. 56, no. 5, pp. 1770–1778, 2008.

[70] J. J. More, "Generalization of the trust region problem," *Optim. Meth. and Soft.*, vol. 2, no. 3–4, pp. 189–209, 1993.

[71] M. Grant and S. Boyd, *Cvx: Matlab software for disciplined convex programming*, 2003. [Online]. Available: http://cvxr.com/cvx

[72] I. Pólik and T. Terlaky, *Interior point methods for nonlinear optimization*, 2010.

[73] J. F. Sturm, "Using sedumi 1.02, a matlab toolbox for optimization over symmetric cones," *Optim. Meth. Softw*, vol. 11, no. 1, pp. 625–653, 1999.

[74] A. F. Molisch, *Wireless communications*, 2nd Ed. John Wiley & Sons Ltd., Chichester, UK, 2011.

[75] T. Eren, O. Goldenberg, W. Whiteley, Y. R. Yang, A. S. Morse, B. D. Anderson, and P. N. Belhumeur, "Rigidity, computation, and randomization in network localization," in *INFOCOM*, Hong Kong, Mar. 2004, pp. 2673–2684.

[76] G. Oliva, F. Pascucci, S. Panzieri, and R. Setola, "Sensor network localization: Extending trilateration via shadow edges," *IEEE Trans. Autom. Control*, vol. 60, no. 10, pp. 2752–2755, 2015.

[77] A. Sahin, Y. S. Eroglu, I. Guvenc, N. Pala, and M. Yuksel, "Accuracy of aoa-based and rss-based 3d localization for visible light communications," in *IEEE VTC*, Boston, MA, USA, Sep. 2015, pp. 1–5.

[78] B. D. S. Muswieck, J. L. Russi, and M. V. T. Heckler, "Hybrid method uses rss and aoa to establish a low-cost localization system," in *SASE/CASE*, Buenos Aires, Argentina, Aug. 2013, pp. 1–6.

[79] E. Elnahrawy, J. A. Francisco, and R. P. Martin, "Adding angle of arrival modality to basic rss location management techniques," in *ISWPC*, San Juan, PR, USA, Feb. 2007, pp. 1–6.

[80] S. C. Ergen and P. Varaiya, "Tdma scheduling algorithms for wireless sensor networks," *Wireless Networks*, vol. 16, no. 4, pp. 985–997, 2010.

[81] J. F. C. Mota, J. M. F. Xavier, P. M. Q. Aguiar, and M. Püschel, "D-admm: A communication-efficient distributed algorithm for separable optimization," *Wireless Networks*, vol. 61, no. 10, pp. 2718–2723, 2013.

[82] D. Dardari, P. Closas, and P. M. Djuric, "Indoor tracking: Theory, methods, and technologies," *IEEE Trans. Vehic. Technol.*, vol. 64, no. 4, pp. 1263–1278, 2016.

[83] J. P. Beaudeau, M. F. Bugallo, and P. M. Djuric, "Rssi-based multi-target tracking by cooperative agents using fusion of cross-target information," *IEEE Trans. Sign. Process.*, vol. 63, no. 19, pp. 5033–5044, 2015.

[84] E. Masazade, R. Niu, and P. K. Varshney, "Dynamic bit allocation for object tracking in wireless sensor networks," *IEEE Trans. Sign. Process.*, vol. 60, no. 10, pp. 5048–5063, 2012.

[85] S. Tomic, M. Beko, R. Dinis, M. Tuba, and N. Bacanin, "Bayesian methodology for target tracking using rss and aoa measurements," *submitted to Elsevier Ad Hoc Networks*, vol. PP, no. 99, pp. 1–1, 2016.

[86] M. W. Khan, A. H. Kemp, N. Salman, and L. S. Mihaylova, "Tracking of wireless mobile nodes in the presence of unknown path-loss characteristics," in *Fusion*, Washington DC, USA, Jul. 2015, pp. 104–111.

[87] M. W. Khan, N. Salman, A. Ali, A. M. Khan, and A. H. Kemp, "A comparative study of target tracking with Kalman filter, extended Kalman filter and particle filter using received signal strength measurements," in *ICET*, Peshawar, Pakistan, Dec. 2015, pp. 1–6.

[88] B. K. Chalise, Y. D. Zhanga, M. G. Amina, and B. Himed, "Target localization in a multi-static passive radar system through convex optimization," *Elsevier Sign. Process.*, vol. 14, no. 102, pp. 207–215, 2014.

[89] S. Tomic, M. Beko, R. Dinis, and P. Montezuma, "Distributed algorithm for target localization in wireless sensor networks using rss and aoa measurements," *Elsevier Perv. Mobile Comput.*, vol. 37, pp. 63–67, 2017.

[90] S. K. Meghani, M. Asif, and S. Amir, "Localization of wsn node based on time of arrival using ultra wide band spectrum," in *WAMICON*, Cocoa Beach, Florida, USA, Apr. 2012, pp. 1–4.

[91] S. Tomic, M. Beko, R. Dinis, and P. Montezuma, "A closed-form solution for rss/aoa target localization by spherical coordinates conversion," *IEEE Wirel. Commun. Letters*, vol. 5, no. 6, pp. 680–683, 2016.

[92] Y. Zou and K. Chakrabarty, "Distributed mobility management for target tracking in mobile sensor networks," *IEEE Trans. Mobile Comput.*, vol. 6, no. 8, pp. 872–887, 2007.

[93] G. Wang, Y. Li, and M. Jin, "On map-based target tracking using range-only measurements," in *CHINACOM*, Guilin, China, Aug. 2013, pp. 1–6.

[94] J. B. B. Gomes, "*An overview on target tracking using multiple model methods*," MSc Thesis, Universidade Técnica de Lisboa, 2008.

[95] R. G. Brown and P. Y. C. Hwang, *Introduction to random signals and applied kalman filtering*, 3rd Ed. Wiley, NY, USA, 1997.

[96] Y. Bar-Shalom and X. R. Li, *Estimation with applications to tracking and navigation*, 1st Ed. John Wiley & Sons Inc., NY, USA, 2001.

Index

About the Authors

Slavisa Tomic received the M.S. degree in traffic engineering according to the postal traffic and telecommunications study program from University of Novi Sad, Serbia in 2010, and the Ph.D. degree in electrical and computer engineering from University Nova of Lisbon, Portugal in 2017. His research interests include target localization in wireless sensor networks and non-linear and convex optimization.

Marko Beko was born in Belgrade, Serbia, on November 11, 1977. He received the Dipl. Eng. degree from the University of Belgrade, Belgrade, Serbia, in 2001 and the Ph.D. degree in electrical and computer engineering from Instituto Superior Tecnico, Lisbon, Portugal, in 2008. Currently, he is an Associate Professor at the Universidade Lusófona de Humanidades e Tecnologias, Portugal. He is also a Researcher at the UNINOVA, Campus da FCT/UNL, Monte de Caparica, Portugal. His current research interests are in the area of signal processing for wireless communications and nonsmooth and convex optimization. He serves as an Associate Editor for the IEEE Access Journal and Elsevier Journal on Physical Communication. He was awarded a Starting Grant under the Investigador FCT programme of the Portuguese Science and Technology Foundation in 2016. He is the winner of the 2008 IBM Portugal Scientific Award.

Rui Dinis received the Ph.D. degree from Instituto Superior Técnico (IST), Technical University of Lisbon, Portugal, in 2001 and the Habilitation in Telecommunications from Faculdade de Ciências e Tecnologia (FCT), Universidade Nova de Lisboa (UNL), in 2010. From 2001 to 2008 he was a Professor at IST. Currently he is an associated professor at FCT-UNL. During 2003 he was an invited professor at Carleton University, Ottawa, Canada.

He was a researcher at CAPS (Centro de Análise e Processamento de Sinal), IST, from 1992 to 2005 and a researcher at ISR (Instituto de Sistemas e Robótica) from 2005 to 2008. Since 2009 he is a researcher at IT (Instituto de Telecomunicações).

Rui Dinis is editor at IEEE Transactions on Communications (Transmission Systems – Frequency-Domain Processing and Equalization) since 2012 and editor at IEEE Transactions on Vehicular Technology since 2015. He was also a guest editor for Elsevier Physical Communication (Special Issue on Broadband Single-Carrier Transmission Techniques). Rui Dinis was TPC chair of the IEEE ICT'2014. He has been actively involved in several international research projects in the broadband wireless communications area.

Milan Tuba is the Dean of Graduate School of Computer Science and Provost for mathematical and technical sciences at John Naisbitt University of Belgrade. He received B.S. in Mathematics, M.S. in Mathematics, M.S. in Computer Science, M.Ph. in Computer Science, Ph.D. in Computer Science from University of Belgrade and New York University. From 1983 to 1994 he was in the U.S.A. at Vanderbilt University in Nashville and Courant Institute of Mathematical Sciences, New York University and later as Assistant Professor of Electrical Engineering at Cooper Union School of Engineering, New York. From 1994 he was Professor of Computer Science and Director of Computer Center at University of Belgrade and from 2004 also a Professor of Computer Science and Dean of the College of Computer Science, Megatrend University Belgrade. His research interest includes heuristic optimizations applied to computer networks, image processing and combinatorial problems. Prof. Tuba is the author or coauthor of more than 150 scientific papers and coeditor or member of the editorial board or scientific committee of number of scientific journals and conferences. Member of the ACM, IEEE, AMS, SIAM, IFNA.

Nebojsa Bacanin was born in 1983 in Belgrade. In 2015 he completed Ph.D. studies in Computer Science at the Faculty of Mathematics, University of Belgrade. Currently he is an assistant professor and vice dean at the Faculty of Computer Science at 'John Naisbitt' University.

He has held classes in several courses at the undergraduate and graduate studies, from operating systems and computer networks to databases and programming. He has been engaged in the project III-44006 from the program of integrated inter-disciplinary researches of the Ministry of Education, Science and Technological Development of the Republic of Serbia. His scientific contribution extends to the optimization and metaheuristic methods based on the intelligence of swarms (swarm intelligence), both for continuous, and for discrete problems. He has published numerous scientific papers in leading international journals indexed in the SCI list (Thomson Reuters ISI), Scopus and/or MathSciNet, as well as, at international conferences indexed in Web of Science, Scopus, and/or IEEE Xplore. He is a regular reviewer of many of the leading top international journals and he participated in organizing committees of the world's leading conferences.